AA003337

Environmental Division

Core Programming Topic at the 2011 AIChE Spring Meeting and 7th Global Congress on Process Safety

Chicago, Illinois, USA
13 – 17 March 2011

ISBN: 978-1-61782-902-4

Printed from e-media with permission by:

Curran Associates, Inc.
57 Morehouse Lane
Red Hook, NY 12571

CURRAN ASSOCIATES INC. proceedings.com

Some format issues inherent in the e-media version may also appear in this print version.

Copyright© (2011) by AIChE
All rights reserved.

Printed by Curran Associates, Inc. (2011)

For permission requests, please contact AIChE
at the address below.

AIChE
3 Park Avenue
New York, NY 10016-5991

Phone: (203) 702-7660
Fax: (203) 775-5177

www.aiche.org

Additional copies of this publication are available from:

Curran Associates, Inc.
57 Morehouse Lane
Red Hook, NY 12571 USA
Phone: 845-758-0400
Fax: 845-758-2634
Email: curran@proceedings.com
Web: www.proceedings.com

TABLE OF CONTENTS

(6a) Upgrading Low BTU Fuels to Reduce Emissions and Enhance Performance In IC Engines

Natali Y. Ganfer and Marco J. Castaldi, Earth and Environmental Engineering, Columbia University, New York, NY

Landfills are currently the most common system used for the final disposal of solid waste. They are projected to remain the most common system in the forthcoming years. The release of Landfill Gas (LFG), mainly composed by CH4 and CO2 in an average proportion of 50%-50%, to the atmosphere constitutes an important environmental issue as methane is a GHG 20 times more powerful than CO2 and 23% of the methane emissions in the US come from landfills. On the other hand, LFG can be a free source of low BTU fuels. For these two reasons, LFG is in most cases being collected, and it can be used as IC engine fuel in order to generate electricity. However, because of the fluctuations in the composition of the gas, higher emissions of pollutants such as UHC, NOx, and CO are released. Thus the importance of lowering the emissions of these toxic gases whose expulsion is regulated by the EPA. One solution to this issue is the addition of Synthesis Gas (Syngas) -H2 and CO- to the input LFG.

Preliminary data has shown that by supplementing the input gas stream with as little as 5% Syngas, Nox emissions can be reduced up to 40%, CO up to 67%, and UHC up to 80%. Both the amounts of CO and UHC are reduced because the Syngas increases the activity of the fuel due to its higher flame speed; t he Nox formation is mitigated as a consequence of a more stable combustion at lower temperatures. More detailed data needs to be collected for different Syngas, LFG, and total input gas compositions. All the tests are being run in a Honda Engine GC160 with Maximum Power output of 2kW. Nevertheless, emission levels of 4 different engines with bigger capacity-Capstone CR200, Solar Centaur 40, Caterpillar G3516 LE-have been compared. It was observed that the results of the experiments can be extrapolated to an industrial scale as well.

See more of this Session: Thermal Processes for Waste Utilization
See more of this Group/Topical: Topical 9: Waste & Biomass Valorization

(6b) A Technoeconomic Analysis of Biodegradable Polymers Production From CO_2 and Waste Plastics

Michael Hartmann, Maoqi Feng, Thomas Andrews and Eloy Flores III, Chemical Engineering Department, Division of Chemistry and Chemical Engineering, Southwest Research Institute, San Antonio, TX

This paper evaluates the economic viability of manufacturing biodegradable polymers (mostly polycarbonates) from CO_2 and waste plastics. Waste plastics recycling are beneficial to environment and economy as well. Thermal processes have been used for feedstock recycling from waste plastics, such as polyethylene, polypropylene and polystyrene. Olefins are very useful feedstock in industry; however, it's expensive to produce olefins by using conventional hydrocarbons cracking technology. In our simulation, olefins are generated from pyrolysis of waste plastics with thermal energy supplied from a solar concentrator, thus no net CO_2 is produced in the pyrolysis. Biodegradable polymers are then produced from the polymerization of CO_2 and olefins directly in a catalytic reactor. This method provides an alternative method for CO_2 sequestration and utilization. A detailed economic analysis using VMGSim software is presented.

See more of this Session: Thermal Processes for Waste Utilization
See more of this Group/Topical: Topical 9: Waste & Biomass Valorization

(6c) Modeling of Medium Constituents for Optimized Production of Lipase From Niger Seed Oil Cake (Guizotia abyssinica) Using Artificial Neural Networks and Genetic Algorithms

Sarat Babu Imandi, Department of Biotechnology, GITAM Institute of Technology, GITAM University, Visakhapatnam, A.P., India and Hanumantha Rao Garapati, Department of Biotechnology, Anil Neerukonda Institute of Technology and Sciences, Visakhapatnam District , AP, India

A hybrid system of Artificial Neural Networks (ANN) and Genetic Algorithms (GA) was adopted to optimize the medium constituents to enhance the lipase production by *Yarrowia lipolytica* NCIM 3589 in solid state fermentation (SSF) using Niger seed oil cake as the substrate. Different microbial metabolism regulating fermentation variables (initial moisture content of the substrate, glucose, and urea) were studied to construct a '3–5–1' topology of the ANN for identifying the nonlinear relationship between fermentation variables and the enzyme yield. ANN model was subsequently optimized using GA for obtaining maximum lipase production. The usage of ANN–GA hybrid methodology resulted in a significant improvement (25 %) in the lipase yield when compared to preliminary investigations. MATLAB 7 was used for implementing the ANN–GA methodology and the corresponding program was included in the appendix.

See more of this Session: Thermal Processes for Waste Utilization
See more of this Group/Topical: Topical 9: Waste & Biomass Valorization

INFLUENCE OF ACIDS AND ALKALIS ON CELLULOSE PYROLYSIS PATHWAYS AND KINETICS

Salim M. Shaik [(a)], Paul Nicholas Sharratt [(a)] and Reginald B.H. Tan [(a,b)]

(a) Institute of Chemical and Engineering Sciences, Singapore 627883, Singapore,

(b) Department of Chemical and Biomolecular Engineering, The National University of Singapore, Singapore

1. Introduction

Cellulose pyrolysis has been the subject of many studies due to it being the largest component of most biomass sources. A phenomenological description of cellulose pyrolysis was hypothesised by Mamleev et al. [1]. The main thermal degradation pathway is via intermolecular transglycosylation reactions within the glucose monomers of cellulose [1,2]. Apart from transglycosylation, formation of liquid tar from cellulose can also occur via β-elimination. Under this mechanism, volatile acids (e.g. carboxylic acids) formed from the initial cellulose decomposition are able to attack the remaining cellulose as Bronsted acids thus catalysing heterolytic (ring-opening) reactions.

In this current work, we have studied the effects of acids (H_2SO_4, H_3PO_4 , H_3BO_3) and alkalis ($Ca(OH)_2$, NH_4OH) on the yields of levoglucosan and its derivatives (levoglucosenone, 1,4:3,6-dianhydro-α-d-glucopyranose, anhydroglucofuranose, hydroxymethyl furfural, furfural and levulinic acid). A proposed mechanism adapted from previous workers [3,4,5] and based on our current results is illustrated in Figure 1.

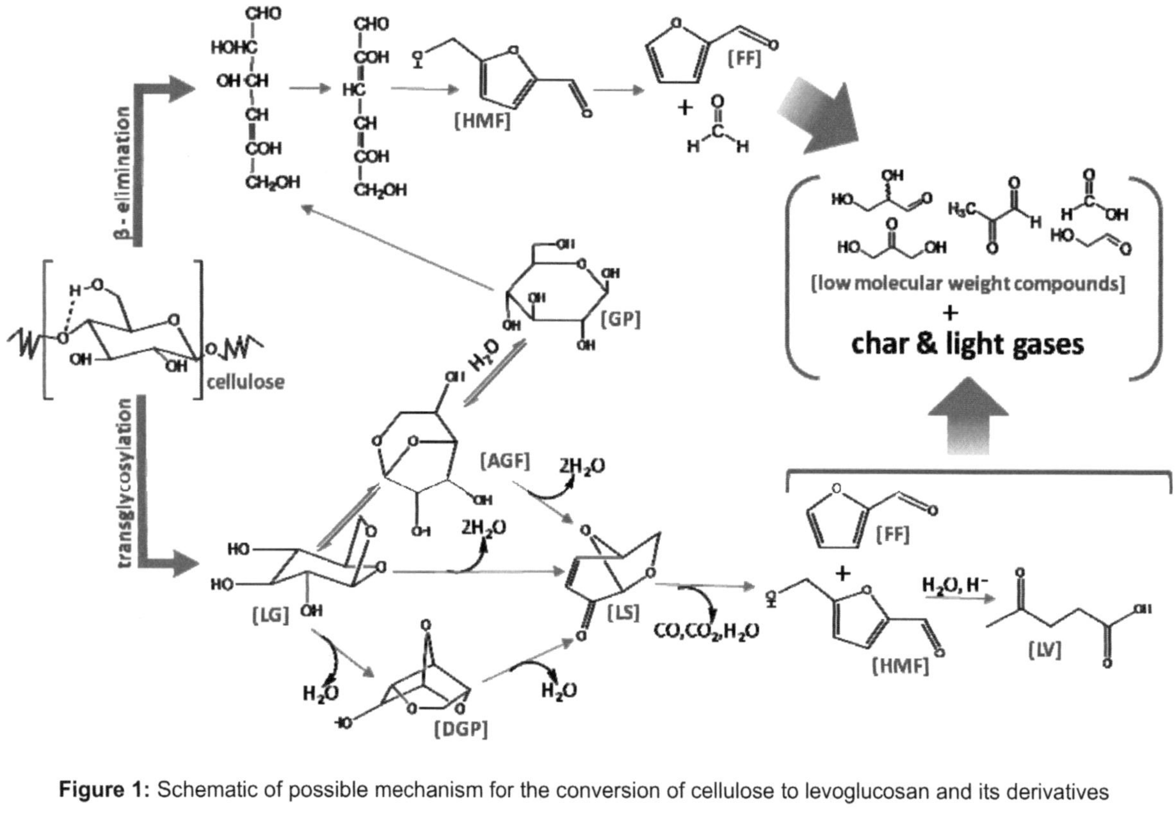

Figure 1: Schematic of possible mechanism for the conversion of cellulose to levoglucosan and its derivatives

In addition to the hypothesised role of volatile acids in pyrolytic chemistry, it has also been known that acid/alkaline content has a significant impact on the yields and product composition during direct liquefaction. Budarin and co-workers had produced bio-oils via microwave pyrolysis of wheat straw [6]. At the same time, they made use of acidic (HCl and H_2SO_4) and alkaline (NH_3) additives to effect changes in the product distribution and composition of the bio-oils derived from wheat straw pellets. The levoglucosan yields of acid and alkaline treated biomass compared with untreated biomass are shown in Table 1 below.

Table 1: Comparison of the effects of acids and alkalis on biomass pyrolysis yields.

	Budarin et al.			Hassan et al.				
additives	H_2SO_4	HCl	NH_3	H_3PO_4	H_2SO_4	NaOH	$Ca(OH)_2$	NH_4OH
levoglucosan	↓ 26%	↑ 27%	↑ 66%	↓ 54%	↓ 61%	↓ 49%	↑ 72%	↑ 41%

Budarin's results were rather mixed with no clear discernable trends between acidic and alkaline additives. A similar trend is also observed based on the results obtained by Hassan et al. [7] who studied bio-oils from pyrolysed pine wood. However, we note that the use of acids (H_2SO_4 and H_3PO_4) decreased levoglucosan yields whilst the alkalis ($Ca(OH)_2$ and NH_4OH) seem to increase levoglucosan yield. Superficially, this seems to provide some evidence for Mamleev's mechanism which suggests that acid catalysis would activate β-elimination whilst the presence of alkalis should suppress β-elimination and promote transglycosylation.

Mamleev's scheme is a novel representation of the reactions involved in cellulose pyrolysis. Although it shares some common features with other models like the Broido-Shafizadeh and Shafizadeh-Chin models, Mamleev seems to be the first to explicitly incorporate the role of acid catalysis in cellulose pyrolysis. Mamleev's scheme, indicated that there is an acid-catalysed (β-elimination) pathway with a higher activation energy (Ea ≈ 250 kJ/mol) compared with the transglycosylation pathway (Ea ≈ 200 kJ/mol) during cellulose pyrolysis. Based on kinetics work done by Banyasz et al. [8] and Capart et al. [9], it was shown that the acid-catalysed, (β-elimination) pathway was favoured as temperature increases (illustrated in Figure 2).

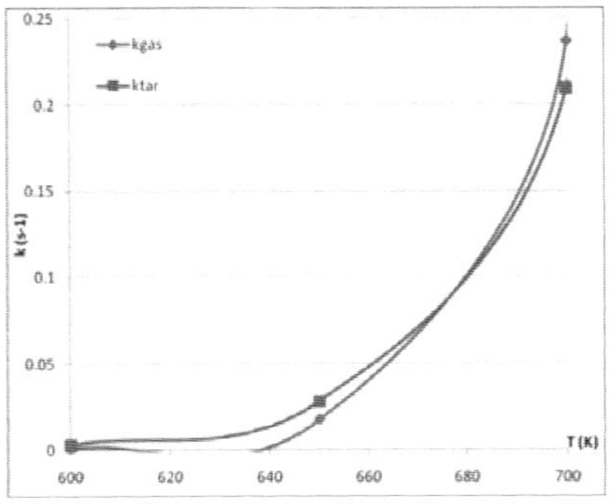

Figure 2a: Comparison of reaction rates for β-elimination (k_{gas}) and transglycosylation (k_{tar}).

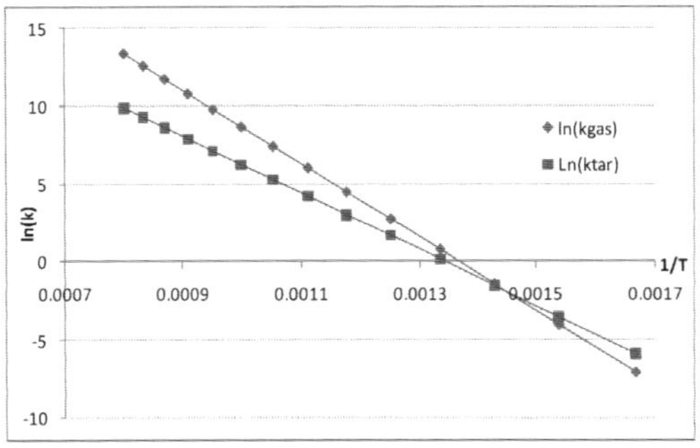

Figure 2b: Comparison of ln(k) against reciprocal temperature.

The graphs in Figure 2 also indicates that the "cross-over" temperature from transglycosylation dominated pyrolysis to ring-opening (β-elimination) dominated pyrolysis occurs at circa 670 K.

The formation of levoglucosan is expected to occur solely via the transglycosylation mechanism. It is therefore expected that the yield of levoglucosan would decrease with increasing temperature and acidity.

2. Experimental Method

The experimental setup used to study levoglucosan yield from cellulose pyrolysis consists of a 1.5 inch diameter stainless steel tubular reactor with a sample holder (capacity ca. 15cm^3) placed inside a horizontal furnace (Carbolite). The furnace is capable of heating the reactor up to 1200 °C and at a rate of 20 to 30 °C/min. Sweep and/or reacting gas is provided via two gas flow controllers which enable variations in gas compositions and reactor residence times. For the current work, nitrogen flow rates circa 650 ml/min were used and this translates to a residence time of 2 to 3 seconds.

Volatile tars and gasses that leave the reaction zone are allowed to cool immediately via three condensers (cooled to ca. 20°C, 0°C and -20°C). The condensed volatiles were analysed using a GC-MS (Agilent: 7890A GC, Column: DB-1701, 60 m x 0.25 mm I.D., 0.25 µm) for levoglucosan and anhydrosacharides content.

Cellulose samples were prepared at different acid and alkali loadings by stirring microcrystalline cellulose (Avicel from Sigma-Aldrich) with the required acid/alkali solution (0.5 to 5wt%) at ca. 25°C for 30 minutes. The "wet" cellulose samples were then filtered and dried overnight in an oven at 60°C.

3. Results and Discussion

The various components of interest namely levoglucosan **(LG)** and its derivatives (levoglucosenone **(LS)**, 1,4:3,6-dianhydro-α-D-glucopyranose **(DGP)**, 1,6-anhydro-β-D-glucofuranose **(AGF)** were identified via their mass spectrum (NIST library) and their retention times The relevant data for the standards are listed in Table 2. Quantification of the various species was done with naphthalene as the internal standard.

Table 2: Identification parameters for the main pyrolysis products

Component	**LG**			**LS**			**DGP**			**AGF**		
Retention time (min)	22.9			17.5			18.8			24.6		
Main mass fragments (m/z)	60	57	73	98	96	39	69	29	57	73	69	44

From the chromatograms, the yields of levoglucosan and its anhydrosacharide derivatives obtained from pyrolysing cellulose with varying acid/alkali loadings were calculated and compared as shown in Figures 3a to 3e.

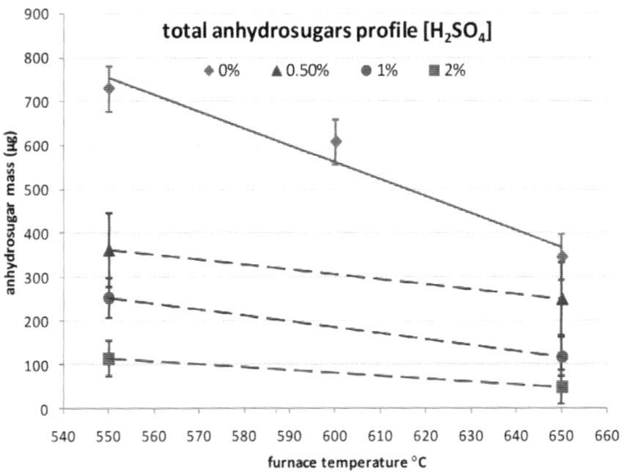

Figure 3a: Anhydrosugar yields from cellulose treated with varying concentrations of H_2SO_4 (0.5wt%, 1wt% and 2wt%)

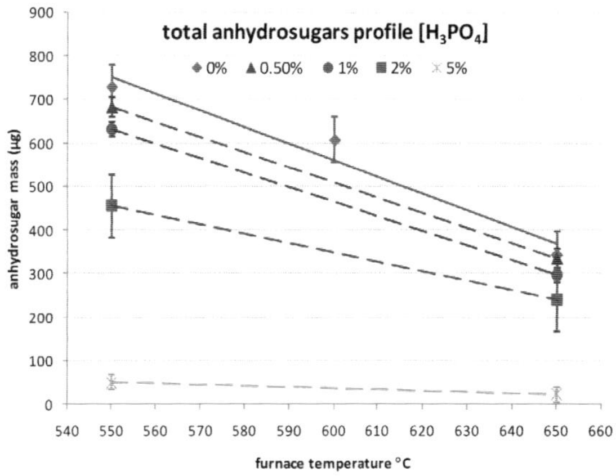

Figure 3b: Anhydrosugar yields from cellulose treated with varying concentrations of H_3PO_4 (0.5wt%, 1wt%, 2wt% and 5wt%)

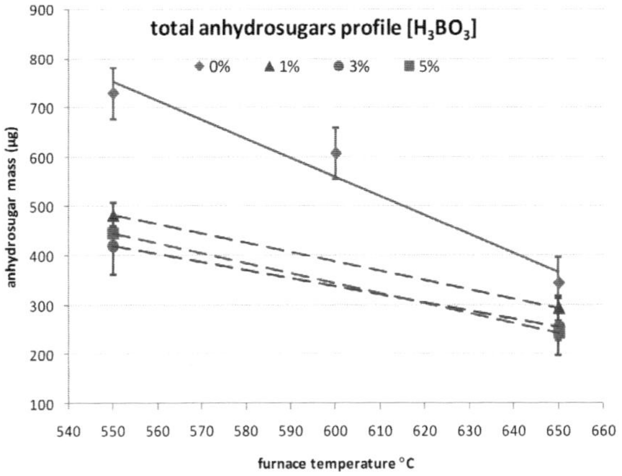

Figure 3c: Anhydrosugar yields from cellulose treated with varying concentrations of H_3BO_3 (1wt%, 3wt% and 5wt%)

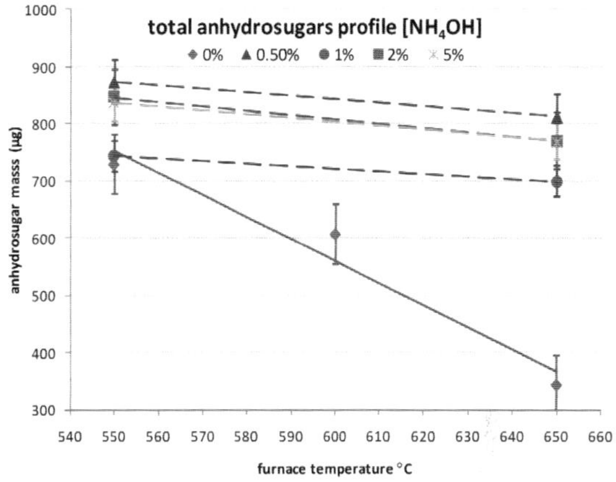

Figure 3d: Anhydrosugar yields from cellulose treated with varying concentrations of NH_4OH (0.5wt%, 1wt%, 2wt% and 5wt%)

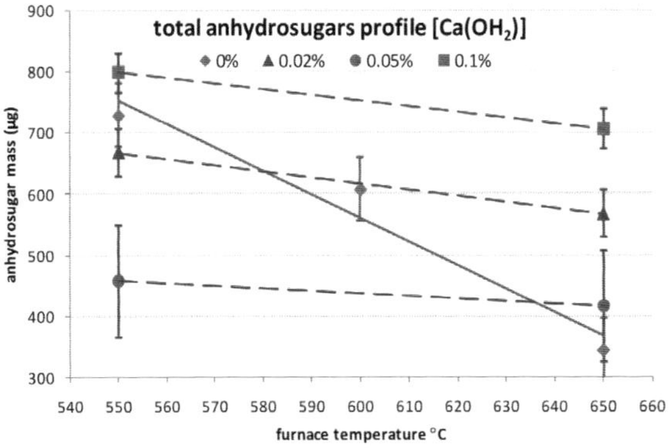

Figure 3e: Anhydrosugar yields from cellulose treated with varying concentrations of $Ca(OH)_2$ (0.02wt%, 0.05wt%, and 0.1wt%)

Looking at the graphs above, we can see that the amount of anhydrosacharides (levoglucosan, levoglucosenone, 1,4:3,6-dianhydro-α-D-glucopyranose, 1,6-anhydro-β-D-glucofuranose) decreases as the reactor temperature is increased. This would be in line with our expectations based on the relative reactions rates for transglycosylation and β-elimination. We can also see that the three acid treatments led to decreased anhydrosacharide yields with higher acid concentrations generally resulting in a corresponding reduction in anhydrosacharides.

The results obtained for cellulose pyrolysis with alkalis were slightly mixed. In the case of ammonium hydroxide, the trend was rather clear that the alkali treatment led to increased levels of anhydrosacharides being produced. However, with the calcium hydroxide treatment, higher levels of anhydrosacharides (compared to untreated cellulose) were only observed at the higher temperatures. In fact, even for ammonium hydroxide, its effects on increasing anhydrosacharides yields were also more pronounced at higher temperatures.

In order to gain further understanding from the results that we have obtained, the relative amounts of levoglucosan, levoglucosenone, 1,4:3,6-dianhydro-α-D-glucopyranose, 1,6-anhydro-β-D-glucofuranose were analysed and is represented in Figure 4.

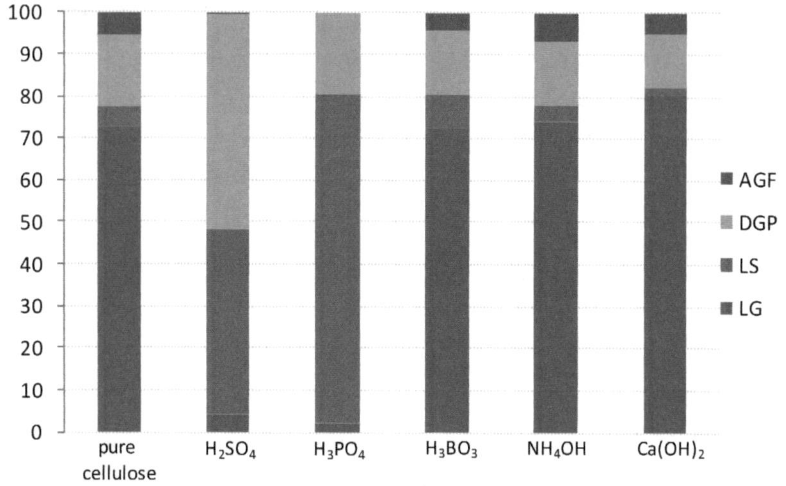

Figure 4: Comparison of the relative yields of the anhydrosacharides [levoglucosan **(LG)**, levoglucosenone **(LS)**, 1,4:3,6-dianhydro-α-D-glucopyranose **(DGP)**, 1,6-anhydro-β-D-glucofuranose **(AGF)**]obtained for the various acid and alkali-treatments.

Figure 4 clearly indicates that treatments with sulphuric and phosphoric acids resulted in higher levoglucosenone levels whereas for boric acid, ammonium hydroxide and calcium hydroxide, the relative amount of levoglucosan was found to be higher.

Based on our results thus far, we can ask the question whether a lower anhydrosacharide yield obtained in the presence of acids was due to the further breakdown of the anhydrosacharides or was it due to promotion of the acid catalysed β-elimination pathway as seen in Figure 1. If we take a closer look at the boric acid treated cellulose runs, the relative amount of levoglucosan was higher compared to the amount of levoglucosenone. This observation can be explained based on the findings of Kawamoto et al. [10] which showed that boric acid forms a stable complex with levoglucosan that prevents further transformation/breakdown. Hence in spite of this, we were still able to observe a decrease in total anhydrosacharide yield for boric acid treated cellulose. This therefore means that the cellulose probably depolymerised via the alternative β-elimination pathway.

On the other hand, we have also seen that at higher temperatures, the alkali-treated cellulose was able to maintain its anhydrosacharide yields better than untreated cellulose. This "stabilising" effect of the alkalis could be due to neutralisation of H^+ ions that are needed to catalyse the β-elimination pathway. This, in turn, allowed the transglycosylation pathway to be favoured during the pyrolysis.

4. Conclusion

Based on these experimental results, it can be seen that the presence of an acid or an alkali can have a significant effect on the yield of anhydrosacharides during pyrolysis. We have observed that acids (H_2SO_4 and H_3PO_4) decreased the total amount of anhydrosacharides and levoglucosan. This is probably due to the acids catalysing the hetrolytic reactions and promoting secondary dehydration reactions which converts levoglucosan into levoglucosenone and 1,4:3,6-dianhydro-α-d-glucopyranose. We have also seen that when the cellulose was treated with alkalis (NH_4OH and $Ca(OH)_2$), the overall yield of anhydrosacharides and levoglucosan increased accordingly.

These two main observations indicate that the β-elimination pyrolysis pathway is likely to be catalysed by the presence of acidic species at the expense of the alternative transglycosylation pathway. This is further confirmed when the presence of an alkali "inhibited" the β-elimination route in favour of transglycosylation.

References

1. **Mamleev V., Bourbigot S., Le Bras M., Yvon J.**, The facts and hypotheses relating to the phenomenological model of cellulose pyrolysis. Interdependence of the steps, *Journal of Analytical and Applied Pyrolysis*, 84, 1-17. 2009.
2. **Graham R.G., Bergougnou M.A.**, Fast pyrolysis of biomass, *Journal of Analytical and Applied Pyrolysis*, 6, 95-135. 1984.
3. **Kawamoto et. al.,** Catalytic pyrolysis of cellulose in sulfolane with some acidic catalysts, *Journal of Wood Science*, 53, 127-133. 2007.
4. **Shen D.K., Gu S.**, The mechanism for thermal decomposition of cellulose and its main products, *Bioresource Technology*, 100, 6496-6504. 2009.
5. **Lin Y.C. et. al.**, Kinetics and mechanism of cellulose pyrolysis, *Journal of Physical Chemistry C*, 113 (46), 20097–20107. 2009.
6. **Budarin et. al.**, The preparation of high grade bio-oils through the controlled, low temperature microwave activation of wheat straw, *Bioresource Technology*, 100, 6064-6068. 2009.
7. **Hassan E.M., Steele P.H.**, Characterization of fast pyrolysis bio-oils produced from pretreated pine wood, *Applied Biochemistry and Biotechnology*, 154, 182-192. 2009.
8. **Banyasz J.L., Li S., Lyons-Hart J., Shafer K.H.**, Gas evolution and the mechanism of cellulose pyrolysis, *Fuel*, 80, 1757-1763. 2001.
9. **Capart R., Khezami L., Burnham A.K.**, Assessment of various kinetic models for the pyrolysis of a microgranular cellulose, *Thermochimica Acta*, 417, 79-89. 2004.
10. **Kawamoto et. al.**, Stable complex formation with boric acid in pyrolysis of levoglucosan in acidic media, *Journal of Analytical and Applied Pyrolysis*, 82, 78-82. 2008.

Organosolv revisited – biomass valorization with new perspectives

Michael Schrems, Philipp Vejdovszky, Martin, Zeilinger, Sonja Bednarik, Emmerich Haimer, Falk Liebner, Thomas Rosenau, Antje Potthast
Christian-Doppler-Laboratory "Advanced cellulose chemistry and analytics"
Department of Chemistry, University of Natural Resources and Life Sciences, A-1190 Vienna, Austria
thomas.rosenau@boku.ac.at, antje.potthast@boku.ac.at

ABSTRACT

This study reinvestigated the well known Organosolv concept aiming at maximum lignin output using $scCO_2$ as co-solvent. Lignin is a still underestimated natural resource with potential application in the near future. The use of N-modified lignins as soil improvers was considered being one of the most interesting utilizations in response to progressive desertification in parts of the world. The presented process uses rye straw as a renewable source achieving up to 70% delignification with lignin purities of 80-85% and nitrogen contents of up to 4%. The nitrogen enrichment was achieved by thermal treatment of rye straw under alkaline Organosolv conditions simultaneously using aqueous NH_4OH as nitrogen source and supercritical carbon dioxide as pulping additive. The effectiveness of the used supercritical carbon dioxide could be demonstrated with a model substance study. Apocynol was used as a representative lignin model showing a significantly faster degradation in the presence of $scCO_2$ which is in accordance to a 40% lower Arrhenius activation energy.

I. INTRODUCTION

Due to fading fossil resources, valorization of renewable resources is on the rise.[1] The enormous demand for energy currently covered through petroleum refinery is the main driving force behind this process. The ongoing change to inexhaustible energy sources does not only require new technologies for biomass processing but will also lead to new industry branches for platform chemicals. Bioconversion of renewable resources by treatment with hot aqueous ethanol, known as the Organosolv process, is an old concept dating back to 1931.[2] Although the idea of separating or extracting components out of wooden materials arose already in the early 19th century, the middle of the 20th century can be considered as the time, when this concept drew attention to a larger community. With the beginning of the 1990s the interest in this research area increased remarkable focusing on the delignification of renewable materials, mainly wood. With the establishment as a promising approach for the future, new possible facets appeared in scientific databases. Organosolv pulping as a conceivable „green" alternative for established sulfur-containing processes as kraft and sulfite, generating enormous amounts of non-recyclable by-products, gained widespread interest with its peak in the nineties. In the following years the focus was directed to other possible renewable raw materials for Organosolv processes such as straw and annual plants. The latest development in this area is the use of Organosolv principles as a pretreatment step for various renewable resources to enhance the bioethanol output from polysaccharides in bacterial digestion.[3] Considerable pretreatment effects can also be obtained with simply using hot water but waive delignification to further increase the enzymatic accessibility.[4,5]

Organosolv conditions involve usually a mixture of water and an aliphatic alcohol, mainly ethanol in various mixtures. Comparable results towards yield can be achieved using methanol although higher purity hence delignification of the pulp can be achieved with ethanol.[6] Use of supercritical carbon dioxide as an additional solvent enables further improvements. Super critical carbon dioxide causes an anomalous porosity and lamellar structure within the material resulting in a higher accessibility for the ethanol water mixture.[7,8] The significantly higher solubility of carbon dioxide in ethanol in opposite to water makes the application of an alcohol water mixture even more preferable over pure hot water pretreatment.[9] Pretreatment steps with similar gases such as sulfur dioxide are also capable to enhance the achievable results but have to be considered less applicable because of their inability to reduce the carbon dioxide foot print.[10]

The everlasting issue in energy production based on renewable materials is how to make such a process profitable. Resources like straw are cheap but the energy input for the transformation to biogas is high rendering a simple conversion less feasible. As pulp resulting from Organosolv treatment of straw and other annual plants does not show sufficient quality to enter the paper and fiber industry, other main constituents need to be transformed to value-added products to sustain such a concept. Lignin, one of the main components in all naturally grown materials, shows a high potential to meet this demand although efficient usage could not be established in the past. Until now rather neglected by simply using it as fuel, various more beneficial applications are possible. The usage of Organosolv lignin as a polyphenol source is just one economically interesting alternative, provided pure isolation.

A very promising application is the use of Organosolv lignin as a source of artificial humic substances which can be applied for recultivation of degraded soils or general soil improvement. Progressive desertification in parts of the world renders this product class highly beneficial in the future. Ammonoxidation is one of the technological opportunities for converting lignin and other ligneous biomaterials into nitrogen-rich organo-mineral fertilizers that share a couple of similarities with natural humic substances. The special value of these modified lignins is the slow release of nitrogen due to different binding forms of nitrogen generated upon ammonoxidation. The fact, that substantially delignified renewable materials are in addition much easier converted to degradation products such as bioethanol and biogas, led us to the development of an organosolv-based bioconversion process maximizing the isolated amount of pure lignin. This can be achieved by applying high pressure ideally with carbon dioxide. As the subsequent ammonoxidation is usually executed in aqueous media, a significant amount of energy can be saved by avoiding the intermediate purification of the lignin. Therefore we combined these two process steps in one by performing the N-modification already during the Organosolv process replacing the water by aqueous ammonium hydroxide.

Understanding the delignification process and the effect of supercritical carbon dioxide on a molecular level is considered being crucial for the process design. First investigations towards the process occurring during organosolv pulping were done in the early 2000s by Kishimoto and Sano.[11, 12] In this work further studies were conducted to clarify the effect of super critical carbon dioxide on the delignification process. For this purpose 4-(1-Hydroxy-ethyl)-2-methoxy-phenol (Apocynol) was used as a representative lignin model substance.

II. EXPERIMENTAL

Chemicals

Chemicals were obtained from commercial sources and were of the highest purity available. Rye straw was obtained from different suppliers in Lower Austria and milled to <1mm containing an average water content of 6.3%. Apocynol was obtained via reduction of acetoguaiacone using sodiumborohydride in 52% yield.

Experimental setup

CO_2 assisted Organosolv pulping was performed in a 0.2 *l* autoclave with jacketed heater (SEPAREX, France). 12 *g* of the $scCO_2$-extraction material were mixed with 100 *ml* 50% (v/v) ethanol in water (initial phase) or aqueous ammonium hydroxide (0.1, 2.5, 5.0 and 10.0 w% NH_3, see **Table 1.**) and equilibrated for 15 minutes. After equilibration the mixture was filled into the preheated autoclave (80 *°C*) and pressurized with CO_2 to 6 *MPa*. Subsequently the heating jacket was heated to the desired temperature. After reaching the adjusted value, the system was pressurized using a hand-piston-pump (Sitec, Switzerland) and left under these conditions until the individual reaction time has elapsed. After the treatment the system was depressurized and the solid residue was washed with 400 *ml* 0.1*M* NaOH and subsequently filtrated. The residue was supposed to be enriched cellulose whereas the supernatant was supposed to bear lignin and hemicellulose. In order to separate these fractions, the supernatant was acidified with HCl to a pH lower than 3 leading to precipitation of the high molecular lignin. After filtration the supernatant as well as the two solid fractions were dried at 70 *°C* and subjected to further analysis.

General Analytics

GPC System. For GPC measurements, the system as described by Röhrling et al[5] was used: online degasser, Dionex DG-2410, Kontron 420 pump; pulse damper; autosampler, HP series 1090; column oven, refractive index (RI) detector, Shodex RI-101. DMAc/LiCl (0.9%, w/v), filtered through a 0.02 *μm* filter, was used as the eluant. The sample was injected automatically, chromatographed on two serial GPC columns, and monitored by RI detection. Molecular weight distribution and related polymer relevant parameters were calculated by software programs. The following parameters were used in the GPC measurements: flow, 1.00 *ml/min*; columns, two, PL gel mixed A LS, 20 *μm*, 7.5x300 *mm*; injection volume,: 100 *μl*; run time, 45 *min*. Data evaluation for lignin molecular weight are based on calibration wit narrow polystyrene standards (Astra 5 software). Standards used for calibration: 30 *kg/mol*, 114.2 *kg/mol*, 200 *kg/mol*, 300 *kg/mol*, 700 *kg/mol*.

Curie-point pyrolysis GC/MS was performed using a CPP-40 pyrolyser (GSG) coupled with a GC 6890 and MSD 5973 (AGILENT TECHNOLOGIES). 260 *μg* of the sample was pyrolysed at 600 *°C* (FECRALLOY™) for 10 seconds. The pyrolysate was carried by helium into the inlet (250 *°C*, split 1:20) of the gas chromatograph. Separation was achieved using a fused silica HP-5ms column (30 *m*, 0.25 *mm*, 25 *μm*) and a column flow of 1.0 $ml\ min^{-1}$. Oven program: 50 *°C* (5 *min*), 5 $°C\ min^{-1}$ to 180 *°C*, than 10 $°C\ min^{-1}$ to 280 *°C* (9 *min*). Auxiliary temperature: 250 *°C* (18 *min*), 10 $°C\ min^{-1}$ to 280 *°C* (14 *min*). Mass spectrometer: EI mode (70 *eV*); ion source 230 *°C*; quadrupole 150 *°C*, $7.6 \cdot 10^{-6}$ *Torr*. Peak assignment was accomplished using the NIST 2005 mass spectra database. Purity evaluation is based on calibration with lignin standards to which 2.5, 5.5 and 10.5 wt.-% of pulp were added. The evaluation of the polysaccharide content of the analysed samples was done by comparing the peaks of furfural and guajacole in the pyrograms. Furfural is a reliable indicator for the sugar content whereas guajacole is the main lignin fragment occurring after thermal fragmentation. Analysis of the hydroxymethyl-furfural peak as main product of hexoses showed that it had only minor effect on the overall result and was therefore not taken into account.

Elemental analysis (C, H, N) was performed at the Microanalytical Laboratory of the University of Vienna.

Analysis of ash and minerals was performed by the Institute of Waste Management at the University of Natural Resources and Life Sciences Vienna. The samples were heated up in aluminium oxide crucibles to 950 *°C* with a rate of 10 *K/min*, gas flow 150 *ml/min* (80% He/20% O_2) using 16mg material per analysis.

HPLC analysis was performed with a Dionex system (P580 pump, GINA 50 autosampler,) running with Chromeleon 4 software. Detector: UVD340U diode detector. Column: Agilent done with a Agilent Eclipse XDB-C18, Injection volume: 10 *µl*, Flow: 0.8 *ml/min*, Gradient program: 0% eluent A→ 95% eluent A (0 *min* → 15 *min*), 95% eluent A (15 *min* → 18 *min*), 0% eluent A (18 *min* → 20 *min*), Detection wavelength: 280 *nm*; Eluent A: 100% MeCN; eluent B: 100% Water.

Statistics

Statistical experiment design was done with CADEMO (Computer Aided Design of Experiments and Modeling) from BioMath GmbH using a Box-Behnken design. Design and calculation were done by Department of Applied Statistics at the University of Natural resources and Life Sciences Vienna.

Synthesis of apocynol

To a solution of acetoguaiacone (7 *g*, 42 *mmol*) in 212 *ml* 56% ethanol 3.5 *g* (93 *mmol*) of $NaBH_4$ are added in portions and afterwards stirred for 10 *h* at room temperature. Another portion of $NaBH_4$ (1 *g*, 26 *mmol*) is added and stirring is continued for additional 14 *h*. The reaction mixture is neutralized with CO_2 (dry ice!), introduced through a glass capillary with a wide diameter to prevent plugging, diluted with 200 *ml* water and extracted six times with 150 *ml* Et_2O. Combined organic phases were dried over Na_2SO_4 (about 6 *h*) and constricted to 200 *ml* under reduced pressure. The ethereal residue is transferred into a crystallization dish and recrystallized when all solvent evaporated. This step is necessary to prevent formation of dimers. The crystallized product is solved in small amounts of ethyl acetate (the unsolvable dimers are filtered off) and precipitated with hexane. The obtained white crystals are dried under reduced pressure. The remaining solution can be constricted and the yellowish crystals can be recrystallized again in the same matter to higher the yield (3,7 *g*; 52%).

^{13}C NMR (75 MHz, $CDCl_3$): 146.61 (*C*-3), 144.98 (*C*-4), 137.93 (*C*-1), 118.32 (*C*-6), 114.15 (*C*-5), 107.99 (*C*-2), 70.33 (*C*-α), 56.00 (OMe), 25.09 (*C*-β)

III. RESULTS AND DISCUSSION

The first part of this study will present investigations towards the Organosolv process. In the initial phase various chemicals e.g. acids and bases have been tested in different concentrations as additives to potentially raise the lignin output and or the purity. Alterations in the consumption of the washing solutions were conducted for the same purpose. After this phase a statistical approach was used to find optimal values for the process parameters. The second part will reveal new insights into the delignification process on a molecular level. Experiments have been done under high pressure at various temperatures. The influence of super critical carbon dioxide was investigated to understand this enhancing factor in more detail.

III.1 Development of an Organosolv process aiming at maximal lignin output

The Organosolv pulping process is known to work best with equivalent alcohol water ratios at temperatures at or slightly above 180 *°C* when high lignin output is desired. The reaction time appears to be directly proportional to the yield leading to increasing yields with increasing time.[3] With this knowledge the test parameters for the initial phase were chosen to be 180 *°C*, 16 *MPa* and 160 *min*. The rather high pressure is due to several factors. Carbon dioxide is known to change into the supercritical state about 31 *°C* and 7.3 *MPa*. More important is the requirement to keep water and also ethanol in the liquid state at this temperature. Additionally high pressures guarantee a high concentration of carbon dioxide in the ethanol-water mixture.

The first tests showed that ethanol gives only slightly better yields than methanol with comparable purity of the obtained lignin fraction. As methanol is approximately twenty percent cheaper than ethanol we continued the study with methanol as the co-solvent. Addition of acid (H_2SO_4) gave good yields for the lignin but very low yields for the cellulose fraction. As this fraction is also relevant for the economics of this concept, this additive has to be considered as unfavorable. The addition of base (NaOH) turned out to be similar unfavorable as this caused very impure lignin fractions. The effectivity of carbon dioxide was proven with two experiments under atmospheric pressure and at 160bar using nitrogen instead of $scCO_2$. The obtained yields were 38% and 13%

lower, respectively. These results indicate that the high pressure itself causes increasing yields but only carbon dioxide can raise the yield even further. Experiments with the washing solution showed that it is not possible to extract lignin with pure water. Very good results could be achieved with 1 *N* NaOH but this led to very high salt concentrations and impurities in the lignin fractions. The usage of 0.1 *N* NaOH gave just 6% lower yields but the salinity and the purity were significantly better. 1 *N* NH_4OH also showed some potential as washing solution but gave 60% lower yields rendering this alternative not very suitable. The determination of the lignin molecular weight M_w was done with GPC analysis. Due to the strong fluorescence of the samples it was not possible to determine the molecular weight with light scattering. Therefore the column was calibrated with polystyrene standards ranging from 30-700 *kg/mol*. The obtained M_w values from the analyzed lignin fractions were relatively homogenous with an average of 7.9 *kg/mol* ± 1.3. Analysis of the ash and mineral content displayed a rather low concentration within the lignin fraction with average ash content of 3.1 % ± 1.5. The next step was to find the optimal operating parameters for the described process. For this purpose a statistical approach as used to reduce the overall number of experiments and avoid systematic errors. The limits for an effective process were considered to be between 150-180 *°C* with pressures between 60-160 *bar*, reactions times between 10-160 *min* and ammonium hydroxide concentrations between 0.1 and 10%. **Table 1** gives an overview about these experiments including achieved yields and purity for the cellulose and the lignin fraction. All purity values have been determined by Curie-point pyrolysis GC/MS.

The bold black highlighted tests 8 and 22 in **Table 1** turned out to be the best pulping conditions when aiming for maximal lignin output. With a straw lignin content of about 20% both conditions lead to delignification of approximately 70% with a lignin purity of 80-85%. These results also show that there is a strong dependency towards all reaction parameters. Depending on the situation it can be chosen between a shorter but higher energy inputs or longer less intensive conditions. With average nitrogen enrichment of 4% in the lignin fraction this process can be considered as a promising concept for the production of nitrogen-rich lignins the latter being worth to become subjects of a thorough study evaluating their fertilizing and soil improving effects as well as their bio- and phytoactivity.

***Table 1:** Statistical experiment design after Box-Behnken*

	T [°*C*]	p [*bar*]	t [*min*]	NH_4OH conc. [%]	Residual cellulosic material [%]	Polysaccharide content pulp [%]	Obtained Lignin [%]	Polysaccharide content lignin [%]
	Block points							
1	157	85	47	2.5	72.3	60.9	1.8	4.7
2	172	85	47	2.5	68.8	60.8	7.7	*n.a.*
3	157	135	47	2.5	75.0	57.5	8.0	*n.a.*
4	172	135	47	2.5	66.1	57.6	4.5	9.0
5	157	85	122	2.5	67.9	55.7	12.5	*n.a.*
6	172	85	122	2.5	59.8	56.7	4.5	12.9
7	157	135	122	2.5	70.5	53.7	2.7	*n.a.*
8	**172**	**135**	**122**	**2.5**	**65.2**	**55.8**	**14.3**	**19.7**
9	157	85	47	7.5	77.7	64.9	8.9	15.8
10	172	85	47	7.5	67.9	64.6	5.4	10.3
11	157	135	47	7.5	59.8	62.4	2.7	8.0
12	172	135	47	7.5	71.4	67.5	11.6	20.3
13	157	85	122	7.5	66.1	72.3	3.6	9.4
14	172	85	122	7.5	72.3	61.3	10.7	11.1
15	157	135	122	7.5	74.1	63.6	9.8	*n.a.*
16	172	135	122	7.5	52.7	*n.a.*	5.4	10.8
	Face points							
17	150	110	85	5.0	62.2	61.6	6.3	9.8
18	180	110	85	5.0	58.9	56.7	7.1	6.8
19	165	60	85	5.0	63.4	57.3	11.6	9.8
20	165	160	85	5.0	59.8	62.0	7.1	*n.a.*
21	165	110	10	5.0	74.1	52.3	3.6	*n.a.*
22	**165**	**110**	**160**	**5.0**	**86.6**	**68.0**	**14.3**	**14.2**
23	165	110	85	0.1	64.3	*n.a.*	8.0	*n.a.*
24	165	110	85	10.0	67.9	55.8	7.1	*n.a.*
	Center point							
25	165	110	85	5.0	67.0	67.6	8.9	*n.a.*

III.2 Investigation of the reaction kinetics through model substance studies

The aim of this study was to obtain a deeper understanding of the reactions occurring during the Organosolv pulping process in the presence of carbon dioxide. The focus was placed on the delignification process in dependence on the temperature applied and the role of supercritical carbon dioxide. For this purpose apocynol seemed to be a suitable model substance for the investigation of the delignification process. It is easily accessible through reduction of acetoguaiacone and representative due to its structure similar to lignin monomer units. [13]

The experiments were conducted in a metal reactor built for high pressure experiments which was pre-heated to 80 *°C*. 20 *mg* of apocynol (**1**) was placed in a small beaker and dissolved in 10 *ml* of a 1:1 mixture of ethanol and water. The beaker with the solution was placed in the reactor and the reaction chamber sealed afterwards. An initial CO_2 pressure of 6 *MPa* was applied to ensure just liquid phases already before reaching the adjusted temperature. The end temperature (140-180 *°C*) was reached within several minutes followed by adjusting a pressure of 16 *MPa* (t=0). At the end of the individual reaction time the heat control was turned off and the metal sheath was cooled with ice to accelerate cooling. Eventually, the orange coloured liquid obtained was analysed by HPLC.

Experiments conducted at 140 *°C*, 160 *°C* and 180 *°C* with and without supercritical carbon dioxide showed that there is an increase in degradation of the model substance with increasing temperature (cf. **Fig. 1**). But this increase is only of minor relevance compared with a much stronger degradation effect in presence of super critical carbon dioxide. As indicated in figure 1, the degradation rate at 140 *°C* with $scCO_2$ present is even faster than the reaction rate at 180 *°C* without $scCO_2$. The order of this reaction was estimated as first order. The calculation of the Arrhenius activation energy gave a significantly lower activation energy for this reaction when $scCO_2$ was applied. An E_A of 68.9 *kJ/mol* for the reaction without $scCO_2$ in comparison to 42.0 *kJ/mol* points out that the usage of $scCO_2$ can lower the activation energy of approx. 40%.

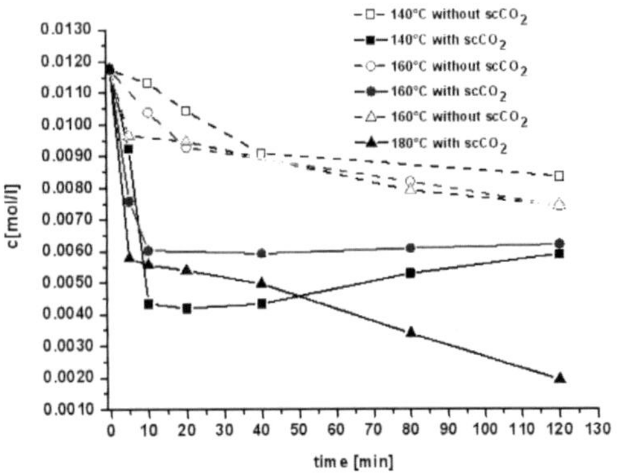

Fig. 1: Degradation of apocynol at different temperatures with and without $scCO_2$

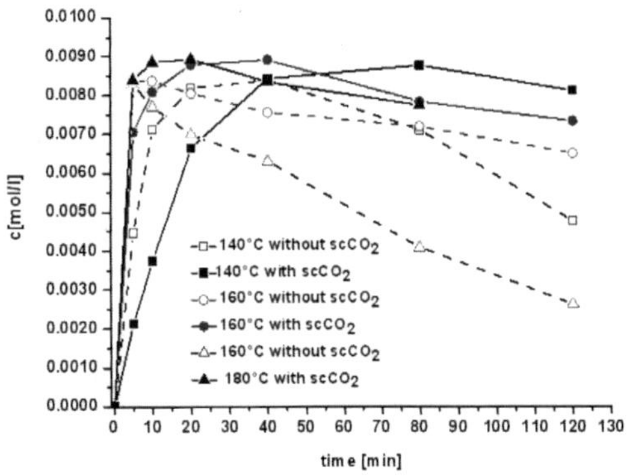

Fig. 2: Formation and transformation of the reaction intermediate 2-methoxy-4-vinylphenol

HPLC analysis of the samples revealed that the overall degradation reaction of the model substance apocynol (**1**) is a combination of two consecutive reactions. The first step is the degradation of the apocynol is the transformation into 2-methoxy-4-vinylphenol (**2**), formed via a *para*-chinone methide (*p*QM), which stabilizes into (**2**) by a [1,7]-sigmatropic proton shift.[14] The second step is the formation of 4-(1-ethoxyethyl)-2-methoxy-phenol (3) by addition of ethanol to the vinyl intermediate according to a pseudo-first-order rate law, cf. **Fig. 3**.

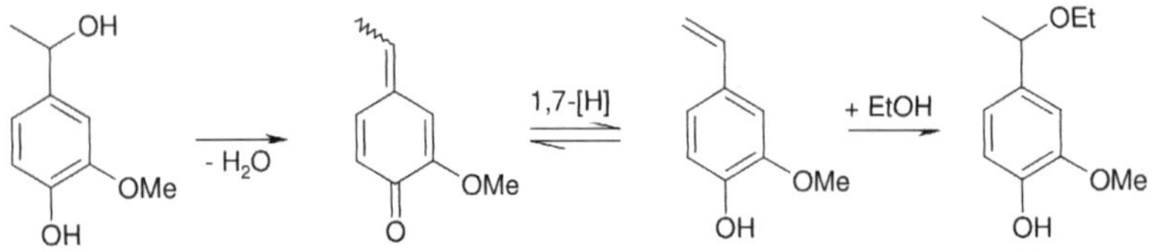

1 **pQM** **2** **3**

Fig. 3: Reaction pathway of apocynol under Organosolv pulping conditions in aqueous ethanol (50%, v/v)

The presence of $scCO_2$ slows down the subsequent transformation of the vinyl intermediate (**2**) into 4-(1-ethoxyethyl)-2-methoxy-phenol (**3**), cf. **Fig. 2**. Whereas in the absence of CO_2 the intermediate (**2**) is transformed within minutes into the final product (**3**), the presence of $scCO_2$ causes a very fast formation of the stable intermediate (**2**) but a decelerated transformation into the end product. Even the effect of temperature almost vanished. The presence of $scCO_2$ also effects the zero order formation of the end product 4-(1-ethoxyethyl)-2-methoxy-phenol (**3**). Although the final concentration is higher when $scCO_2$ is applied the initial reaction rate is definitely higher when no $scCO_2$ is present. This finding is in accordance with the calculated values for the Arrhenius activation energy. The E_A for this reaction without $scCO_2$ was determined with 4.6 *J/mol* which is about four to five times lower than the calculated value of 20.9 *J/mol* for $scCO_2$.

IV. CONCLUSIONS

The developed process allows isolating major parts of the total lignin content in rye straw. Significant increase of the lignin output can be achieved by applying high pressure during the process. Highest amounts can be obtained when $scCO_2$ is used during the reaction. This effect can be explained by the super critical state of the carbon dioxide at the investigated reaction conditions making the raw material better accessible and acting as a solvent enhancer. A statistical approach was used to find the optimal regions for all operating parameters. Temperature, pressure and time had a significant influence on the results. Additional tests were done to evaluate the contributions of various additives and auxiliaries namely acids and bases and the possible exchange of ethanol for methanol which can save up to twenty percent in the overall solvents costs.

The study showed that it is possible to achieve a delignification of approximately 70% with a sufficiently pure lignin fraction. Furthermore it was proven that nitrogen enrichment, conventionally performed subsequently, can be accomplished during the pulping process significantly reducing energy demand. A nitrogen enrichment of 4% within the lignin fraction is considered to be sufficient for the aimed purpose of soil improvement.

Kinetic studies clearly showed the influence of supercritical carbon dioxide on the reaction mechanism. Apocynol was used as a simple model for lignin. Degradation kinetics with and without super critical carbon dioxide were recorded at different temperatures between five minutes and two hours. The decay of the original model substance and the formation of degradation products were analyzed by HPLC. The use of super critical carbon dioxide during the process does not only accelerate the decay of the model substance but also lowers the activation energy distinctly, which is in accordance with current literature. It was shown that the presence of $scCO_2$ only accelerated the transformation into the intermediate product (**2**) but slowed down the formation of the end product from that intermediate. It seems quite clear that the application of super critical carbon dioxide has a positive impact on the delignification of the designed process. Tests replacing the $scCO_2$ with nitrogen within the process and the accelerated transformation of the model compound apocynol confirmed the expectations of a supportive effect. The hindrance of $scCO_2$ on the formation of the apparently end product 4-(1-ethoxyethyl)-2-methoxy-phenol can be explained with a potentially different reaction environment making the intermediate compound more stable. To get a better insight in the role of $scCO_2$ on the delignification process, a study using a dimeric compound containing a β-O-4 ether bond similar to the one used in experiments of Kishimoto and Sano is planned.[11]

V. ACKNOWLEDGEMENT

The financial support by the Christian Doppler Research Society (CD-laboratory "Advanced Cellulose Chemistry and Analytics") and Blue Globe Energy, Mannheim, Germany, is gratefully acknowledged. The authors would like to thank Prof. Dr. Ena Smidt from the Institute of Waste Management at the University of Natural Resources and Life Sciences Vienna for performing the ash and minerals content analyse and Prof. Dr. Karl Moder from the Department of Applied Statistics at the University of Natural resources and Life Sciences Vienna for the statistical experimental design.

VI. REFERENCES

1. J. H. Clark, V. Budarin, F. E. I. Deswarte, A. J. Wilson, Green chemistry and the biorefinery: a partnership for a sustainable future, *Green Chem.* **2006**, *8*, 853-860.

2. T.. Kleinert, K. Tayenthal, Ueber neuere Versuche zur Trennung von Cellulose und Inkrusten verschiedener Hölzer, *Angewandte Chemie* **1931**, *44*, 788-791.

3. X. Pan, N. Gilkes, J. Kadla, K. Pye, S. Saka, D. Gregg, K. Ehara, D. Xie, D. Lam, J. Saddler, Bioconversion of Hybrid Poplar to Ethanol and Co-Products Using an Organosolv Fractionation Process: optimization of Process Yields, *Biotechnology and Bioengineering* **2006**, *94(5)*, 851-861.

4. M. Zeng, N. S. Mosier, C-P. Huang, D. M. Sherman, M. R. Ladisch, Microscopic Examination of Changes of Plant Cell Structure in Corn Stover Due to Hot Water Pretreatment and Enzymatic Hydrolysis, *Biotechnology and Bioengineering* **2007**, *97(2)*, 265-278.

5. K. Öhgren, R. Bura, J. Saddler, G. Zacchini, Effect of hemicellulose and lignin removal on enzymatic hydrolysis of steam pretreated corn stover, *Bioresource Technology* **2007**, *98*, 2503-2510.

6. M. Oliet, J. Garcia, F. Rodriguez, M. A. Gilarranz, , Solvent effects in autocatalyzed alcohol-water pulping – Comparative study between ethanol and methanol as delignifying agents, *Chem. Eng. J.* **2002**, *87*, 157-162.

7. M. Gao, F. Xu, S. Li, X. Ji, S. Chen, D. Zhang, Effect of SC-CO_2 pretreatment in increasing rice straw biomass conversion, *Biosystems Engineering* **2010**, *106*, 470-475.

8. D. Pasquini, M. T. B. Pimeta, L. H. Ferreira, Sugar cane bagasse pulping using supercritical CO_2 associated with co-solvent 1-butanol/water, Journal of Supercritical Fluids **2005**, *34*, 125-131.

9. E. Dalmolin, E. Skovroinski, A. Biasi, M. L. Corazza, C. Dariva, O. Olivieira, Solubility of carbon dioxide in binary and ternary mixtures with ethanol and water, *Fluid Phase Equilibria* **2006**, *245*, 193-200.

10. P. Sassner, M. Galbe, G. Zacchi, Steam Pretreatment of *Salix* with and without SO_2 Impregnation for Production of Bioethanol, *Applied Biochemistry and Biotechnology* **2005**, *121-124*, 1101-1118.

11. T. Kishimoto, Y. Sano, Delignification Mechanism during High-Boiling Solvent Pulping. Part 1. Reaction of Guaiacylglycerol Ether, *Holzforschung* **2001**, *55*, 611-616.

12. T. Kishimoto, Y. Sano, Delignification Mechanism during High-Boiling Solvent Pulping. Part 2. Homolysis of Guaiacylglycerol-β-Guaiacyl Ether, *Holzforschung* **2002**, *56*, 623-631.

13. E. Adler, S. Delin, G. E. Miksche, Synthese des Isoeugenolglykol-β-(2-methoxyphenyl)-äthers, *Acta Chem. Scand.* **1966**, 20(4), 1035-1037.

14. D. Fengel, G. Wegener, Wood: chemistry, ultra structure, reactions, deGruyter, Berlin, New York, **1989**.

This welcoming session sponsored by Fuels & Petrochemicals Division (F&PD) is a new initiative organized by the AIChE's national and F&PD technical programming team as a preamble to the division's sponsored technical sessions. This will be opened by a short introduction by the chair of the division. Besides recognizing leadership in his introductory short presentation, the chair will highlight the key elements of value additions the division brings to its members. The presentation will be concluded with a special request to join the division where professional opportunities are limited only by members' imagination, to positively impact academia and industry and thereby society at large. For details of the division and its many facets of activities and functions please follow the link www.aiche-fpd.org.

(10b) What Are We Going to Do with All This Natural Gas? - Ramifications of the Dramatic Expansion of Natural Gas Available to the Market Due to North American Shale Gas Production and LNG Liquefaction

Sam Thomas, Chart Industries, Inc., Garfield Heights, OH

The success in Shale and other unconventional formations has dramatically expanded the N. American Natural Gas reserves and promises to also do so in many other parts of the world. New LNG liquefaction capacity in the Middle East and Asia along with additional large scale capacity from Australia and Papua New Guinea coming on-line in the next 5 years opens the opportunity for a global trade in Natural Gas stabilizing historically volatile pricing. How will this important energy and petrochemical feedstock be utilized? The success in improving techniques to allow unconventional gas reserves to be proven and significantly increase the available reserves has not as yet been fully appreciated by markets and policy makers. This presentation reviews the ramifications of natural gas becoming a larger part of the global energy supply for which there are numerous competing interests. Natural Gas holds the ability to significantly reduce carbon emissions by substitution for coal in power generation, it can be used as a transportation fuel either as LNG, CNG or GTL clean diesel increasing energy independence of some of the largest economies in the world simultaneously reducing carbon emissions. It could enable the USA to maintain its position as a leading petrochemical producer if used for ethylene feedstock. How will these competing interests play out and how will governments respond?

See more of this Session: Fuels & Petrochemicals Division Welcome Session and Energy Leadership Presentation
See more of this Group/Topical: Topical 7: 14th Topical on Refinery Processing

(14a) The Energy Challenge: Providing Sustainable Supplies to Meet Growing Demand

F. Emil Jacobs, ExxonMobil Research and Engineering Company, Annandale, NJ

As economies grow and living standards improve, the world faces a dual energy challenge: meeting the growing demand for energy while mitigating emissions and their impact on the environment. ExxonMobil believes that addressing this dual challenge requires an integrated set of solutions which includes increasing the supply of all commercially viable energy sources, reducing emissions, and improving energy efficiency in all sectors of the economy. Technology will be a key enabler of continued progress in all of these areas.

Many experts agree that the immense scale of global energy demand necessitates a continuing role for oil and gas as a substantial portion of the energy supply mix for at least the next several decades. These experts also forecast that the overall energy mix will shift toward a relatively higher level of sustainability through a combination of increased efficiency across the economy, growth in lower-carbon sources of energy, and development and deployment of technologies that help mitigate emissions.

ExxonMobil has developed a number of technologies that contribute to increased supplies of energy, reduced emissions, and higher efficiency in its operations and in consumer use of its products. The presentation will provide an overview of ongoing research and development activity in a broad range of technology areas including algae-based biofuels. It will also highlight some of company's current technologies including lithium-ion battery separator films, lightweight plastics that improve automotive efficiency, and processes that could advance carbon capture and storage.

See more of this Session: Spring Meeting Keynote Address - Sponsored by the AIChE Fuels & Petrochemicals Division
See more of this Group/Topical: Liaison Functions

COMMERCIAL SCALE TRITIUM RECOVERY FROM EXIT SIGNS AND OTHER ILLUMINATION DEVICES AT LLNL

Diane K. Spencer, J. Mark Mintz, Matthew G. Lucas and Dale M. Holck

LLNL-CONF-467413

Lawrence Livermore National Laboratory
Livermore, CA 94551
spencer2@llnl.gov, mintz1@llnl.gov, lucas3@llnl.gov

Abstract

The disposal of exit signs and similar commercially available tritium-powered illumination devices in landfills can pose potential public health and environment risks. These risks can significantly be reduced by collecting the gaseous tritium remaining in these devices. Since 2001, Lawrence Livermore National Laboratory (LLNL) and the U.S. Army have been engaged in a successful tritium recovery and recycle program for a limited number of specialized devices. This laboratory-scale work has been accomplished by hand with reasonable efficiency. However, recent requests for assistance with similar work include millions of these devices and would require much larger capabilities. Thus, LLNL has installed a commercial-scale tritium recovery system—the Tritium Grinder System (TGS)—in its Tritium Facility to complement its laboratory-scale tritium recovery capabilities. The TGS was built to efficiently process tritium-bearing illumination devices on a large scale. It can process up to 72 standard exit signs in a batch, with significantly more for smaller illumination devices (e.g., 2250 rotary phone dials per batch). The TGS liberates gaseous tritium from these various devices by mechanically grinding (or shredding) them under negative pressure in a modified industrial shredder. The liberated tritium is collected in a storage tank and periodically removed by trapping on a regenerable hydrogen getter. This paper presents the results from initial grinder operations, including the collected gas composition and unit processing rate, along with a summary of the design features, operating history, and safety controls.

Overview

Studies of active landfills show leachate with tritium considerably above background, with the tritium concentration in many above the Environmental Protection Agency drinking water MCL of 20,000 pCi/L (Ref. 1). Thus, the Nuclear Regulatory Commission (NRC) regulates tritium-containing illumination devices in general and exit signs in particular requiring reporting and handling requirements, including proper disposal of unwanted or unused signs. The requirements in most cases prevent disposal as ordinary trash. LLNL has recognized that the residual tritium in these discarded devices is valuable and has developed a method to capture it.

TGS Design and Process

The “grinder” portion of the TGS is a commercially available shredder unit from SSI Shredding Systems, Inc., modified to operate under negative pressure. A batch of devices is delivered to the grinder by a conveyor belt and is mechanically shredded. The released tritium, along with decay product helium-3 (3He), and argon (Ar) added as a sweep gas, is then pumped to a collection tank. The collected gas is processed through a commercially available getter to remove the tritium. Solid debris (e.g., exit sign housing residuals) generated through the recovery process is collected in a lined 55-gallon drum using a “bagout” process engineered to minimize outgassing and worker exposure. Upon completion of the bagout process, the content of the drum is disposed of as low-level radioactive waste.

Loading, Grinding, and Tritium Collection

Figure 1 shows the major components of the grinding and collection portion of the TGS. Illumination devices to be processed are introduced through the access door and loaded onto the conveyer belt. Small devices (e.g., telephone dials) are placed inside baskets. After the conveyor is loaded, the access door is closed and the system is pumped down. Argon purging is employed to minimize residual air and water vapor. When the system reaches a suitable negative pressure, the illumination devices and baskets are shredded, with the debris accumulating in the discharge hopper. Strategically positioned cameras allow operators to follow the progress of the baskets up the conveyor and into the grinding chamber. Evolved gas is pumped through a particulate filter to the collection tank. A series of purges with argon is then used to ensure a more complete transfer of the tritium gas from the grinder system.

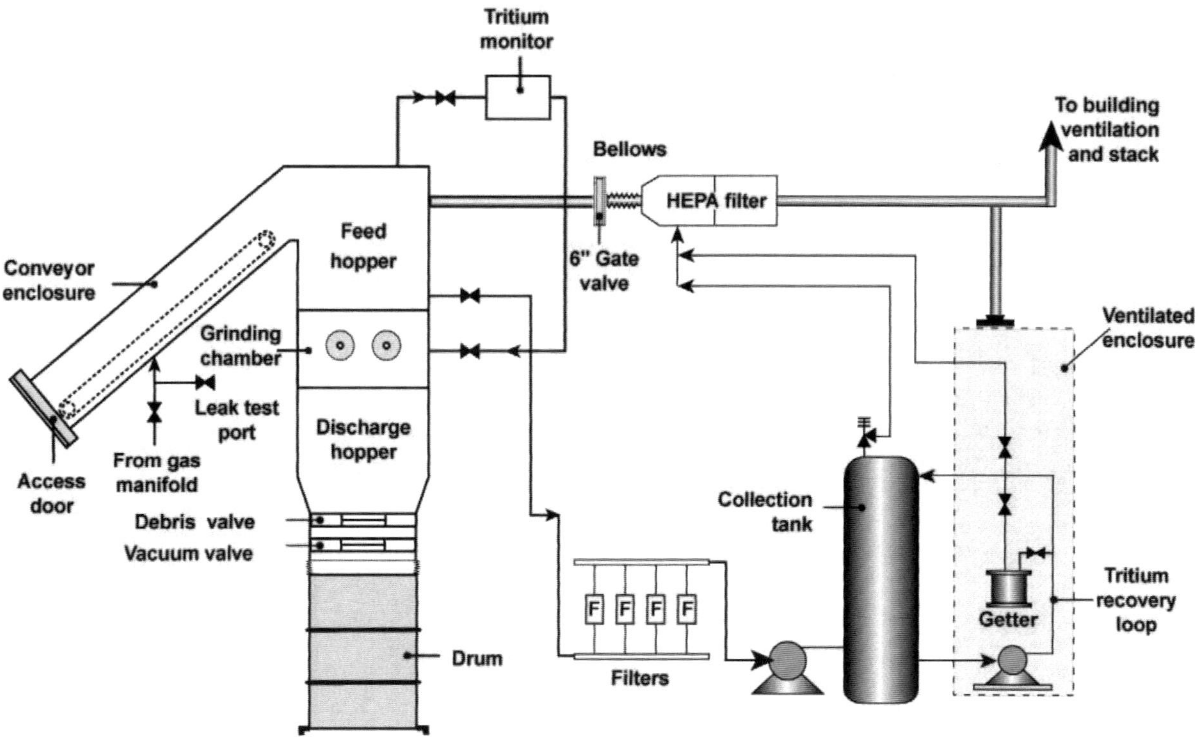

Figure 1. Major components of the TGS.

Following the final argon purge and pump out, the grinder enclosure is let up to ambient pressure, then exhausted to the building ventilation to permit tritium sampling, disposal of debris, and conveyor reloading. A dedicated tritium monitor is consulted prior to venting to verify that less than 1 Ci of tritium is released to the building ventilation when the grinder system is opened. A high efficiency particulate air (HEPA) filter protects the building ventilation from possible particulate formed during the shredding process and, in turn, the building ventilation provides airflow to protect workers from residual contamination.

The grinder is equipped with a double valve near its base. The vacuum valve is relied upon to contain tritium during processing. The debris valve above the vacuum valve supports the accumulated shredded material.

The solid debris generated during the recovery process is safely collected in a lined 55-gallon drum using a "bagout" technique similar to that developed for actinide gloveboxes. This technique allows for the waste drum liners to be sealed and replaced without external exposure. Periodically, a video camera and a small vacuum cleaner nozzle are inserted through a small slit in the liner to inspect and clean the vital vacuum valve seal. The slit is then tape sealed when the camera is removed. Following closure of the drum liner, the filled low-level waste drum is set aside for disposition, and new drum is set in its place.

Tritium Purification, Storage, and Disposition

The central feature of the tritium recovery loop is a high-capacity tritium getter (SAES CapaciTorr-B, MK5 Type). The recovery loop is used to remove tritium (and other active gases) when the pressure in the collection tank reaches 700 torr or 600 Ci, whichever is reached first. The relatively pure argon gas remaining in the collection tank after gettering is exhausted to the building ventilation, but the possibility of recycling that gas is under consideration.

Operating Experience

Grinder startup operations are conducted in two trial period phases using obsolete British Telecom rotary dial telephones used from 1964 to 1981. Phase 1 began on August 11, 2010, and concluded on September 15, 2010. During this phase, approximately 17,250 dials were processed. Phase 2 operations are ongoing, and include collecting tritium on the getter. Phase 2 is expected to be completed by April 2011. Both phases are designed to gain sufficient experience to understand the hazards associated with grinding operations and develop and implement the required controls.

Recovered Tritium

The dials processed consist of a tritium-filled, horseshoe-shaped glass tube glued onto an 84-mm diameter ABS plastic backing plate. The United Kingdom estimated that each dial contains approximately 40 millicuries (mCi) of tritium, with approximately 5 mCi as tritium gas and 35 mCi as tritium embedded in the solid matrix; the latter is not tightly bound. Data collected during Phase 1 operations showed that of the 550 Ci of tritium (HT, DT or T_2) associated with the 17,250 phone dials processed, 58 Ci were recovered. This 10.5% recovery rate compares well with the 12.5% expected from earlier measurements in the United Kingdom. The slight discrepancy may be the result of tritium labeled as methane and water known to be present in trace amounts, but not yet accounted for quantitatively.

Production Rate

Based on results to date, TGS operators can comfortably process two batches of four baskets each of low-curie illumination devices such as phone dials in an average "production" day. Operations involving higher curie items such as signs (~5 Ci each) would likely be limited to one batch a day because of a facility requirement to remove tritium from the collection tank when it approaches 600 Ci.

Collected Gas Analysis

The collection tank pressure was 320 torr at the end of Phase 1. Table 1 shows a mass spectrometric analysis of the collected gas; Table 2, the combined mole fractions of the hydrogen species. The vast majority (99%) is the purged gas Ar, followed by air constituents (O_2, N_2, and CO_2) totaling 0.86%. The presence of air is not surprising, though the O_2/N_2 fraction at 4.5, roughly the inverse of air, is interesting and could be due merely to differences in the Ar sweep efficiency.

Turning to the light isotopes, the preponderance of protium (^1H) at 93% of total hydrogen is likely the primary result of tritium exchange with environmental protium present as water vapor in the glass walls of the phone dials tritium-containment tubes (Ref. 2). Protium originating as molecular hydrogen (H_2) within the glass also may contribute to this value. The mole fraction of protium is estimated by adding the mole fractions of H_2 and half the mole fractions of hydrogen tritide (HT) and hydrogen deuteride (HD), yielding 0.0693 mole fraction %. The total hydrogen species is 0.074 mole fraction % (see Table 2).

The presence of deuterium (D2, DT, HD) can be explained as a known impurity in the original fill gas. The amount of deuterium collected is broadly consistent with the known history of these dials. For example, the approximately identical mole fractions of tritium and deuterium are reasonable given that more than 3 half lives of decay ('average' dial filled in 1972) have converted most of the original tritium fill to ^3He. Tritium half life is 12.5 years and decays to ^3He. After 38 years of decay, approximately 12% of the tritium remains. This suggests that an original deuterium fraction on this order is reasonable and well within the significant uncertainties involved in this estimate.

Table 1. Collected Gas Composition.

Compound	Mole Fraction (%)
Ar	99
O2	0.67
N2	0.15
CO2	0.033
H2	0.066
^{3}He	0.026
He	0.0085
HT	0.0035
HD	0.0032
D2	0.00078
T2	0.00045
DT	0.00036
Methanes	trace
Waters	trace
^{38}Ar	trace
Total	**100**

Table 2. Hydrogen Species Mole Fractions.

Compound	Mole Fraction (%)
Hydrogen (H2, HD/2, HT/2)	0.069
Deuterium (D2, HD/2, DT/2)	0.0026
Tritium (T2, HT/2, DT/2)	0.0024
Total Hydrogen Species	**0.074**

The observed ^3He/tritium ratio of 11 is more difficult to interpret quantitatively due to hydrogen/helium permeation rate differences through the phone dial glass containment tubes and other factors. In the absence of permeation or other differential loss mechanisms, however, a $^3He/T_2$ ratio of about 15 would be observed. This result suggests that permeation of ^3He is faster than tritium through the dial glass as would be expected (Ref. 3).

Hazards and Controls

Phase 1 grinder operations were designed primarily to gather information on the nature and type of radiological hazard the recovery process might pose. Very conservative assumptions were made about the nature and characteristics of tritium contamination that might be generated by the grinding process. In particular, the possibility that residual fine particulates incorporating with or otherwise being contaminated with tritium (i.e., Special Tritium Compound) (Ref. 4) could be a hazard to workers was taken into account. Thus, the area where grinder operations are performed is treated as an "airborne contamination area" and a very high degree of worker protection is specified. Worker protection measures include use of hooded Powered Air Personal Respirators (PAPR, 3M Model GVP-100) and tape-sealed gloves and booties, along with extensive airborne particulate and surface swipe monitoring programs.

Tritiated Particulates

Breathing zone air samplers and continuous passive air samplers ("goose neck" monitors) confirmed that airborne particulates were either extremely low in concentration or not present and therefore not a problem for grinder operations with phone dials. The breathing zone samplers have been discontinued for subsequent operations, but the continuous passive air samplers are being used in the Phase 2 trial period. Thus, the posting in the highest work area was downgraded from "Airborne Contamination Area" to "Contamination Area."

Spreadable Contamination

Consistent with past experience with manual processing of U.S. Army illumination devices, moderate levels of surface contamination (2000–5000 dpm/100 cm^2) were found in several working areas. Light decontamination (mopping) between grinding sessions has proven effective in keeping this spreadable contamination under control.

Personnel Protective Equipment

For Phase 2, use of a full-face HEPA respirator is required only when the conveyor access door is open or while performing bagout-related activities.

Path Forward

The operational trial period for TGS is expected to end April 2011. After that, exit signs and other illuminating devices as well as telephone dials will be processed routinely. Helium-3 is a valuable isotope for fusion research and other applications; the facility is exploring the feasibility of also capturing this material from obsolete illumination devices.

Acknowledgments

This work performed under the auspices of the U.S. Department of Energy by Lawrence Livermore National Laboratory under Contract DE-AC52-07NA27344.

References

1. *Lack of Tritium Exit Sign Control and Contamination of Landfill Leachate*, Association of State and Territorial Solid Waste Management Officials; ASTSWMO Radiation Focus Group, Federal Facilities Research Center, Washington, DC, July 2009.
2. *Tritium in Scottish Landfill Sites*, Report prepared for the Scottish environment Protection Agency by Galson Sciences Ltd., May 2000.
3. Shelby, J. E., Handbook of Gas Diffusion in Solids and Melts, ASM International, Materials Park, OH, April 1996.
4. DOE-HDBK-1129-2008, Tritium Handling and Safe Storage, U.S. Department of Energy, Washington, DC., December 2008.

(21b) Behavior Physics-Mechanical Properties of the Cementitious Matrix From Arum,, Vegetal Fiber, Cement, Metacaulnita and Sand

Goreti M. S. M.[1], **Felix Gamarra**[2], Ramildo D. Tolído filho[1] and **Raimundo Vasconcelos**[3], (1)Civil Engineering, Federal University of Rio de Janeiro, Rio Janeiro, Brazil, (2)Department of Gama, University of Brasilia, UnB Gama College, Brasilia, Brazil, (3)Department of Technology, Federal University of Amazonas, Manaus, Brazil

Resent researches during the last three decades showed that the application of natural products have increased with more intensity. A valuable service would therefore be to find materials alternative as fibers to enhanced physic-mechanical properties of cementitious matrices. In this work was evaluated the capability of *Arumã* vegetable fiber with possibilities to reinforce cementitious matrix. The *Arumã (Ischnosiphon polyphyllus),* is found in the Amazon biome. These fibers are widely used by the tradition people of the Northern Region in handicrafts and construction. The objective this work was studied the behavior physic-mechanical of the reinforcement of cementitious matrix from *Arumã* fibers, ciment, metacaulinita and sand. We find that the values of some properties as strength, modulus of elasticity and traction in the cementitious matrix made were compatible with the values of literature and commercial standards. The results showed to highlight potential that exist in this fiber for developing of cementitious matrix

See more of this Session: Non-Thermal Valorization of Waste and Biomass
See more of this Group/Topical: Topical 9: Waste & Biomass Valorization

(21c) Influence of Process Conditions on Properties of Ash From Biomass Gasification

Naomi Klinghoffer, Earth and Environmental Engineering, Columbia University, New York, NY, Marco J. Castaldi, Department of Earth and Environmental Engineering (Henry Krumb School of Mines), Columbia University, New York, NY and Ange Nzihou, Chemical & Environmental Engineering, Ecole des Mines d'Albi-Carmaux, ALBI Cedex 09, France

Influence of process conditions on properties of ash from biomass gasification

Naomi Klinghoffer[1], Marco J. Castaldi[1], Ange Nzihou[2]

[1]Columbia University Department of Earth and Environmental Engineering

[2]Ecole des Mines d'Albi Carmaux, Department of Chemical and Environmental Engineering

Abstract

Biomass is likely to be a significant energy resource in the future. A common way to recover energy from biomass is through gasification where synthesis gas is produced; a by-product of this process is ash. This research investigates the potential to use the ash as a catalyst by understanding the properties of ash generated under different gasification conditions. Specifically, it is desired to produce a porous ash which could be used as a catalyst or as a support for more catalytically active metals.

In this work, poplar wood was gasified under CO_2, steam, and air at different reaction temperatures.

Experiments were done in a fluidized bed reactor at temperatures of 750°C and 920°C and ash was recovered. BET-surface area measurements showed that ash from gasification under CO_2 had a higher surface area than ash produced from steam gasification. TGA experiments showed that gasification under CO_2 resulted in a higher mass loss compared to gasification with steam. Gasification with steam/ CO_2 mixtures yielded a mass loss similar to that of steam only which could be indicative of competitive reactions between steam and CO_2.

Introduction

Gasification is a promising technology to yield a usable gaseous product from biomass or waste with high efficiency. There are three main components that are produced from gasification: light gases (such as H_2, CO, and CO_2) that can be used for synthesis of fuels or combustion applications, tars (heavier hydrocarbons), which need to be removed as they can cause problems with downstream equipment, and ash, which is made up of minerals and metals that do not enter into the gas phase. It has been shown that the characteristics of the ash (for example, composition and porosity) can be affected by the reactive gases that are used for gasification(1). The porosity of the ash, combined with high content of metals and minerals, makes ash a good candidate to be used as a catalyst(2). It is common that ash generated from gasification of biomass or waste is either sent to a landfill or used in concrete. Use of ash in catalytic applications would increase its value as well as avoid the need for more expensive catalysts. This research investigates the potential to use the ash as a catalyst by understanding the properties of ash generated under different gasification conditions. Specifically, it is desired to produce a porous ash which could be used as a catalyst or as a support for more catalytically active metals.

Experimental

In this work poplar wood was gasified and analyzed at three different scales. Gasification was done in an in-house built fluidized bed reactor; various flow-through gases were used and results are presented here from experiments done under 10% CO_2 in N_2 and 90% H_2O in N_2. The reactor was heated up to a maximum temperature of 750°C or 920°C and held at the maximum temperature for 1 hour or 30 minutes. The top of the reactor was equipped with a porous frit therefore all ash was trapped inside the reactor and recovered after experiments. BET surface area of the ash samples was measured (Micrometrics Gemini). TGA/DSC (Setaram 111) experiments were done where the biomass samples were heated to 800°C at 20°C/min Experiments were done at different concentrations of CO_2 and H_2O in N_2. Gasification was also done in an environmental scanning electron microscope (FEI XL30) to observe the changes in physical properties of the sample during gasification. Gasification was done under air, CO_2 and H_2O and images were recorded throughout the experiments.

Results and Discussion

After fluidized bed experiments, the ash was recovered and BET-surface area was measured. Results are shown in Figure 1. When comparing surface area of ash generated under steam at 750°C, it is clear that longer gasification time leads to a higher surface area ash. However, mass recovery data shows that the mass of ash recovered is similar in both cases; for 30 minutes 5.60% of the initial mass is recovered as ash and at 1 hour 4.95% of the initial mass is recovered. TGA tests were also done with steam at 800°C (with 9%H_2O in N_2); after 30 minutes at 800°C the mass loss is 80.5% and after 1 hours mass loss is 82.7%. Therefore, both TGA tests at low H_2O concentrations and fluidized bed experiments at high concentrations of H_2O both show that mass loss is very gradual at temperatures in the range of 750-800°C. However, the BET surface area of the ash from fluidized bed experiments showed very different surface areas based on the time spent at maximum temperature. At 1 hour the surface area of the ash sample is 621m^2/g and at 30 minutes the surface area is 429m^2/g. This indicates that it is possible to produce ash with significantly different properties without necessarily influencing overall production of gas/tars. This is important because if it desired to produce ash with specific properties (for example, higher porosity), one would want to do this without impacting overall gas yield.

A.

□□p>**Figure 1** Ash recovered from fluidized bed experiments: A. surface area B. mass recovery

Gasification was done in the fluidized bed at 750°C for 30 minutes under CO_2 (10% CO_2 in N_2) and H_2O (90% H_2O in N_2) and BET-surface area was measured. The mass recovered was 15.44% under CO_2 and 5.6% under H_2O; it is expected that the mass loss would be higher under steam since more reactant was introduced. However, the surface area of the two samples was the same. This indicates that the CO_2 produces a more porous ash per gram of biomass reacted. Therefore, if it desired to produce a more porous ash, CO_2 is likely a better candidate than H_2O.

TGA tests were done on biomass under 100% CO_2, 9% CO_2 in N_2, 9% H_2O in N_2, and $H_2O/CO_2/N_2$ mix (9%, 8%, balance), as shown in Table 1. The highest mass loss was 98.25% with 100% CO_2. It is reported in the PHYLLIS database that poplar wood has an ash content of 1.5% (dry) therefore the results with pure CO_2 demonstrate almost complete reaction. In the temperature range of 100-200°C, there is dehydration of the sample and a higher mass loss is observed for the dry samples where there is a stronger driving force for water to leave the sample. When 200°C is reached the mass loss under pure CO_2 is 7.11% while the mass loss under 9% CO_2 in N_2 is 5.85%. If dehydration is the only process taking place then one would expect equal mass loss in both cases, since both use dry gases. For example, TGA experiments were done at with air and N_2 at identical conditions (heating rate of 2°C/min to

500°C); both gases were dry. While the final mass loss was very different for the two gases (98.4% for air and 76.6% for N_2), the mass loss was similar until temperatures of 200°C. Therefore, it is possible that under CO_2 reaction begins to take place at temperatures as low as 150°C. ESEM images also showed more physical changes at low temperatures under CO_2 compared to H_2O. For example, at 400°C the pores in the biomass sample have expanded under CO_2 whereas the sample under steam has very similar physical properties to the raw biomass, as shown in Figure 2.

Table 1 Mass loss in TGA gasification experiments

Gas	Final mass loss (%)
100% CO_2	98.25
9% CO_2 in N_2	86.66
9% H_2O in N_2	81.95
9% H_2O, 8% CO_2 in N_2	81.27

Figure 2 ESEM images A. CO_2, raw biomass B. CO_2, 430°C C. H_2O, raw biomass D. H_2O, 425°C

Mass loss was 86.66% under 9% CO_2 in N_2 and 81.95% under 9% H_2O in N_2. The higher mass loss under CO_2 could be partly due to the Boudouard reaction, shown in Equation 1. It was also interesting to note that mass loss under the mixture of CO_2 and H_2O was almost identical than mass loss under H_2O only (81.27% with CO_2 and H_2O and 81.95% with H_2O only). This could be indicative of a competitive reaction between H_2O and CO_2.

Equation 1 $CO_2 + C$ fl $2CO$ *Boudouard Reaction*

Conclusion

If the ash generated from biomass gasification is to be used in catalytic applications it is important to understand the properties of the ash as well as the ash yield that one can expect from different reaction conditions. Overall, it is desirable to produce an ash with catalytic properties without significantly impacting overall gas recovery. Gasification with CO_2 produces a more porous ash compared to gasification under steam. A higher mass loss is observed under CO_2 compared to steam and reactions under CO_2 begin at lower temperatures compared to steam.

References: (1) Butterman, H.C.; Castaldi, M. J. *Environ. Sci. Technol.* **2009**, *43*, 9030-9037

(2) El-Rub, Z. A.; Bramer, E. A.; Brem, G. *Fuel*. **2008**, 87, 2243-2252

See more of this Session: Non-Thermal Valorization of Waste and Biomass
See more of this Group/Topical: Topical 9: Waste & Biomass Valorization

(21d) Bringing Value to Waste Through Gas Fermentation

Michael A. Schultz, LanzaTech Limited, Roselle, IL

Waste gas streams are produced in many process industries, where they are either used as a fuel source, or simply are flared before being released to the atmosphere. Gas streams rich in carbon dioxide, water, carbon monoxide and/or hydrogen also can be produced through combustion or gasification of solid waste streams such as municipal solid waste, forestry residues, or agricultural residues, where the resulting gas stream is used to produce heat and/or power. In both of these cases, the main driver is to reduce the waste stream, or to convert a toxic species such as carbon monoxide to something more benign such as carbon dioxide, as well as to extract value by production of power and/or heat.

Biological fermentation of carbon monoxide rich gas streams provides an alternative route to conversion of these waste streams. The LanzaTech process produces ethanol and other high-value fuel and chemical products from renewable resources that feed on these gas streams. The carbon monoxide rich gas stream can contain any amount of hydrogen, providing an opportunity to process waste streams from a wide range of industries. The brings a significant upgrade in value to the carbon in the waste stream as compared to heat or power production, as well as an opportunity to reuse the carbon rather than release it as CO2 to the atmosphere.

See more of this Session: Non-Thermal Valorization of Waste and Biomass
See more of this Group/Topical: Topical 9: Waste & Biomass Valorization

(21e) Evaluating Different Methods Used to Manufacture Particleboards Produced Using the Blend of Plant Waste From Industrial Processes and Different Resins

Felix Gamarra, Department of Gama, University of Brasilia, UnB Gama College, Brasilia, Brazil, **Valquiria Quirino**, Forestry, Virginia Tech, Blacksburg, VA, Miriam Lima, Department of Technology, Federal University of Amazonas, Brazil, Raimundo Vasconcelos, Department of Technology, Federal University of Amazonas, Manaus and **Edison Bittencourt**, Chemical Engineering, State University of Campinas, Campinas, S,,o Paulo, Brazil, Brazil

Solid wastes from industrial processes have the potential of polluting groundwater, the atmosphere, and soils. This makes its final disposal a major environmental concern. A valuable service would therefore be to find alternative uses for this waste. In this work plant waste generated by the industry was used to produce particleboards. Two types of panels were produced. The first used the blend of waste from Cumaru (*Dipteryx odorata*) and polyurethane resin. The second used the blend of waste from wood of Eucalyptus (*Eucalyptus* sp.), asparagus (*Asparagus officinalis)*, sugar-cane (*Saccharum officinarum*), and urea-formaldehyde (UF) resin. Physical-mechanical properties of the particleboards, including modulus of rupture, strength, modulus of elasticity, and perpendicular traction, were evaluated following the EN and the ASTM norms. The results obtained in this work were satisfactory. The values found were compatible to those found in the literature and in commercial standards. Thus, this experimental work demonstrated the potential use of renewable resource to produce panel for the civil industry.

Keywords: Vegetal waste, particleboard, physical-mechanical properties, resin.

See more of this Session: Non-Thermal Valorization of Waste and Biomass
See more of this Group/Topical: Topical 9: Waste & Biomass Valorization

(21f) Surface Modification of Various Substrates for Microbial Adsorption Optimization

Lee-At Sabbagh, **Diane Hildebrandt**, David Glasser and T. Matambo, Comps, University of Witwatersrand, Johannesburg, South Africa

Conversion of biomass to useful bio-products is a lengthy and often inefficient process. In our research, we have looked at the conversion of cellulose to ethanol by means of termite bacterial consortium in a fluidized bed bioreactor, where the bacteria are attached as bio-film to an activated carbon or cellulose support. Fairly high conversion rates were achieved in this reactor and the process was fairly robust and flexible. However establishing the bio-film on the activated carbon was a lengthy and fairly chaotic process in terms of the time it took the bacteria to attach to the various substrates (Activated Carbon and Cellulose) and form flocculants. The formation of such flocculants substantially increases the reaction surface and hence should optimize the production of ethanol. Bacteria carry a negative surface charge and hence for increased attraction, the surface charge of the substrate should be modified to be positive. This research, performed initially as batch processes, has shown that with the correct surface charge modifications of the substrate the electrostatic attraction forces between the surface and the bacteria are maximized. As a result of the strong electrostatic attraction forces between the two surfaces the bacteria adsorbs and attaches to the substrate quicker and creates a thicker lining on the surface than seen previously. With these results optimization of flock creation in a fluidized bed bioreactor and its effect on the production of ethanol will be further investigated.

See more of this Session: Non-Thermal Valorization of Waste and Biomass
See more of this Group/Topical: Topical 9: Waste & Biomass Valorization

(23a) Dramatic Nano-Fluidic Properties of Carbon Nanotube Membranes and Application In Chemical Processes and Water Treatement

Bruce J. Hinds, Chemical and Materials Eng., University of Kentucky, Lexington KY, KY

Carbon nanotubes have three key attributes that make them of great interest for novel membrane applications 1) atomically flat graphite surface allows for ideal fluid slip boundary conditions 2) the plasma cutting process to open CNTs inherently places functional chemistry at CNT core entrance and 3) CNT are electrically conductive allowing for electrochemical reactions and application of electric fields gradients at CNT tips. To explore the hypothesis of 'Gatekeeper' membrane selectivity, the entrances to CNT's cores were functionalized with aliphatic amines of different lengths, charged dye molecule and poly-peptides and a hindered diffusion model with a geometry of CNT tip functionalization was consistent with the experimentally observed selectivities. Changes in the flux and selectivity also support a model where charged tethered molecules at the tips are drawn into the CNT core at positive bias hindering/gating flux across the membrane. In general, the transport mechanisms through CNT membrane are a) ionic diffusion is near bulk expectation with no enhancement from CNT b) gas flow is enhanced by ~1-2 order of magnitude due to specular reflection off of flat graphitic surface c) and pressure driven flux of a variety of solvents are 4-5 ORDERS OF MAGNITUDE FASTER than conventional Newtonian flow due to atomically flat graphite planes inducing nearly ideal slip conditions. This is a nearly ideal platform to induce electro-osmotic flow with high charge density at pore entrance and a nearly frictionless surface for the propagation of plug flow. Through diazonium electrochemical modification we have successfully bound anionic surface charge to CNT tips and along CNT cores. High electro-osmotic flows of 0.05 cm/s at -300mV bias are seen with 40 fold improvements in power efficiency. Initial application demonstration in programmed drug delivery, catalytic conversion platform, bioseparations and water treatment are highlighted.

See more of this Session: University Research In Water, Bioseparation, and Pharmaceutical Field for Industry
See more of this Group/Topical: Topical 5: University Research for Industry

(23b) Fouling Resistant Membranes

Benny D. Freeman, Chemical Engineering, University of Texas at Austin, Austin, TX

An urgent need in the 21st century is access to secure, sustainable sources of fresh water. Membrane-based desalination has been recognized as an energy efficient, low environmental impact route to produce clean water for a variety of applications. Additionally, ultrafiltration and microfiltration of aqueous streams have become important unit operations in many industries, including the water purification and wastewater purification industries. This presentation focuses on 2 new classes of polymers to attack critical problems in membrane-based water purification: (1) chlorine-tolerant desalination membranes and (2) fouling-resistant membrane nanolayers based on polydopamine.

Commercially available RO membranes are prepared mainly from aromatic polyamides (PA). However, PA membranes suffer from poor resistance to continual exposure to oxidizing agents such as chlorine. Chlorine is the most widely used oxidizing biocide in water treatment because it is inexpensive and highly effective when present in water at levels of a few ppm. Disinfection of feed water to membrane desalination units is required to prevent biofilm growth on the membranes, which significantly degrades their performance. However, PA membranes cannot tolerate continuous exposure to water containing more than a few ppb of chlorine. This presentation discusses results from a systematic study of the desalination properties of a new family of highly chlorine tolerant, sulfonated polysulfones. These polymers are prepared via a novel synthesis involving direct copolymerization of non-sulfonated monomers with a disulfonated monomer. By controlling the content of disulfonated monomer in the final polymer, desalination properties, such as water permeability and salt rejection, can be varied over a wide range. A family of materials, based on random copolymers, phase-separated block copolymers, blends, and crosslinked materials may be prepared from these starting blocks to provide a new platform for desalination membranes. We are currently working with a major producer of desalination membranes to reduce these new materials to practice.

Across many platforms of membranes, fouling mitigation is a major challenge to be addressed to achieve the most energy-efficient, cost-effective membrane filtration processes. Previously, many surface modifications and functionalized polymers were reported to prevent fouling. However, most of these techniques and materials are practically difficult to implement in water purification membranes. We have discovered surface treatment methodologies that can be used to prepare high permeability polymeric membranes from all common water purification membrane classes. These surface-modified membranes have persistent tolerance to fouling by proteins and emulsified oil, two ubiquitous contaminants in a variety of wastewaters. These membranes were prepared by depositing bio-inspired, self-polymerized hydrophilic polydopamine nanolayers on their surfaces. To demonstrate scalability of this modification, the nanolayers were also applied to entire membrane modules. Upon nanolayer deposition, the membranes could be further functionalized using fouling-resistant macromolecules, such as poly(ethylene glycol), to promote improved fouling resistance and, therefore, high membrane flux. The simplicity, versatility, and broad applicability of the nanolayer deposition technique on membranes provide distinct advantages over other modification strategies to reduce fouling. A start-up company, Advanced Hydro, Inc., based in Austin, Texas has been formed to commercialize these approaches to improve membrane fouling resistance.

See more of this Session: University Research In Water, Bioseparation, and Pharmaceutical Field for Industry
See more of this Group/Topical: Topical 5: University Research for Industry

(23c) Synthesis of Reactive Nanoparticles In Membranes

D. Bhattacharyya[1], V. Smuleac[1] and Scott R. Lewis[2], (1)Chemical and Materials Engineering, University of Kentucky, Lexington, KY, (2)Chemical & Materials Engineering, University of Kentucky, Lexington, KY

Nanostructured metals have become an important class of materials in the field of catalysis, optical, electronic, magnetic and biological devices due to the unique physical and chemical properties. The nanosized metals precipitated from solutions are extremely reactive due to the high surface energy, and they usually form aggregates without the protection of their surface. Therefore, immobilization of metal nanoparticles in polymer membrane (such as, cellulose acetate, PVDF, polysulfone, chitosan, etc,) media is important from the point of view of reactivity, organic partitioning, preventing loss of nanoparticles, and reduction of surface passivation. Another major advantage of having a polymer domain is that nanoparticles (without causing agglomeration) can be directly synthesized in the matrix. This presentation will include technology developments relating to direct synthesis (including "green chemistry" approaches) of iron and bimetallic nanoparticles in microfiltration membranes for environmental applications. The project has been funded by NIEHS , US. EPA, and DOE-KRCEE.

See more of this Session: University Research In Water, Bioseparation, and Pharmaceutical Field for Industry
See more of this Group/Topical: Topical 5: University Research for Industry

(23d) The Design of Advanced Functional Membranes for Industrial Bioseparations

Scott M. Husson, Department of Chemical and Biomolecular Engineering, Clemson University, Clemson, SC

Market demand is increasing rapidly for biotherapeutics such as recombinant proteins, monoclonal antibodies, viral vaccines and plasmid DNA. With the advent of molecular biotechnology and engineered cell lines, upstream production processes have made unprecedented progress in the last decade, shifting the production burden to the downstream processing. To improve process economics and meet market demand, manufacturers will require higher productivity and higher resolution separation techniques – membrane chromatography fulfills these requirements.

In this talk, I will describe our efforts to develop adsorptive membranes for use as chromatography media. I will describe the design and performance characterization of these materials and dispel two common misperceptions by showing that (1) membrane chromatography can be a higher capacity process than resin chromatography in the purification of biologics, and (2) membrane chromatography can be a higher resolution process than resin chromatography.

Our design approach is to graft polyelectrolyte nanolayers from the pore surfaces of macroporous membranes using a controlled polymerization method. Using a number of examples, I will demonstrate that the dynamic binding capacities of proteins and DNA can be dramatically higher for well-designed macroporous membrane adsorbers than for existing commercial membrane adsorbers and resin columns. Using controlled polymerization and high polymer chain density produces adsorptive membranes with high binding capacities that are independent of flow rate, enabling high-throughput operation.

See more of this Session: University Research In Water, Bioseparation, and Pharmaceutical Field for Industry
See more of this Group/Topical: Topical 5: University Research for Industry

(23e) Microfluidic Platforms to Screen for Crystallization Conditions of Active Pharmaceutical Ingredients

Sachit Goyal[1], Michael R. Thorson[2], Yuchuan Gong[3], Geoff G.Z. Zhang[3], Charles F. Zukoski[4] and Paul J. A. Kenis[5], (1)Department of Chemical & Biomolecular Engineering, University of Illinois at Urbana-Champaign, Urbana, IL, (2)Department of Chemical & Biomolecular Engineering, University of Illinois at Urbana Champaign, Urbana, IL, (3)Solid State Sciences, Abbott Laboratories, North Chicago, IL, (4)Department of Chemical and Biomolecular Engineering, University of Illinois at Urbana Champaign, Urbana, IL, (5)Chemical & Biomolecular Engineering, University of Illinois at Urbana-Champaign, Urbana, IL

Screening for polymorphs is critical in the API development process. Identifying the available solid forms of candidate drugs (CD) as early as possible require intensive screens varying multiple parameters including counter ions, antisolvent and concentration of the CD.

Presently, highly complex robotic systems are used at the macroscale that utilize sample volumes of the order of 5 μl per condition [1,2]. Due to small quantity of the CD available in the early stages, number of conditions that can be screened is limited. Microfluidics have the potential to improve high throughput screening by utilizing reduced sample volumes, allowing improved control over mixing of fluids via free interface diffusion, and increasing the number of conditions that can be screened simultaneously [3].

Here we present a microfluidic platform enabling high throughput screening (48 wells on one platform) of antisolvents for polymorphism and crystal habit of APIs utilizing extremely small sample volumes (50nl per well). Several parameters like ratio of volume of API to antisolvent (varied from 1:5 to 5:1), solvent, and API concentration are varied to control supersaturation and are found to influence crystal habit and polymorphism. The microfluidic platform has been modified to enable in-situ Raman spectroscopy for the analysis of crystals and validation of polymorphs crystallized on-chip without need to manually harvest crystals from the chip. To support experimental results, we modeled the effect of geometry of wells on the supersaturation profiles and hence on the nucleation kinetics of the API with antisolvent. We demonstrate the capability of the chip using two model APIs – Indomethacin and Naproxen to generate different polymorphs and crystal morphologies on chip.

References

[1] P. Desrosiers, *Modern Drug Discovery*, 2004, 40.

[2] S. Talreja, D.Y. Kim, A.Y. Mirarefi, C.F. Zukoski, P.J.A. Kenis, J. Appl. Crystallography, 2005, 38, 988-995.

[3] C. Hansen, E. Skordalakes, J. Berger, S. Quake, *Proc. Nat'l Acad. Sci.*, 99, 2002, 16531-6.

See more of this Session: University Research In Water, Bioseparation, and Pharmaceutical Field for Industry
See more of this Group/Topical: Topical 5: University Research for Industry

An efficient p-chlorophenol removal prepared from cationic gemini surfactant modified montmorillonite

Xuemin Liu[1,2], Chunxu Lu[2], Cong Song[1], Songying Wang[1], Yi Wang[1]

(1.School of Chemical and Material Engineering, Jiangnan University, Wuxi Jiangsu, 214122, China; 2.School of Chemical Engineering, Nanjing University of Science and Technology, Nanjing Jiangsu, 210094, China)

Corresponding author. E-mail address: liu65xm@yahoo.com.cn

Abstract: A group of Gemini cationic surfactants containing three hydrophobic and two hydrophilic groups were used as the spacer to modify sodium montmorillonite (MMT-Na) via cation exchange to produce a *p*-chlorophenol removal. The Gemini cationic MMT was characterized with FT-IR and XRD, and its thermal stability was analyzed using thermogravimetric analysis (TG).

The modified montmorillonites performed a good removal of *p*-chlorophenol in wastewater. The *p*-chlorophenol adsorption amount of the Gemini cationic MMT is up to 39.36 mg/g, which is 7 times higher than that of MMT-Na. The modified montmorillonites also present a good thermal stability and the decomposition happened at 200-500°C mainly due to initial weight loss from residual water desorption, followed by decomposition of the surfactant and the dehydroxylation of structural water in the montmorillonite layers. The structure-function relationship of the surfactant was also investigated.

Keywords: Gemini cationic surfactant; Montmorillonite; modification; *p*-chlorophenol; adsorption

1. Introduction

Chlorophenols are widely used as pesticides and antiseptics. In addition to commercial production, chlorophenols can be produced when waste or drinking water is disinfected with chlorine. Various chlorophenols with isomers are commonly found organic pollutants in subsurface environments. Direct exposure to chlorophenols can be fatal. Long-term exposure to the chemicals may cause cancers and affect the function of the liver and the immune system. Numerous researchers have investigated methods to remove organic pollutants from aqueous phases (Lawrence et al., 1998; Wibulswas et al., 1999; Juang et al., 2004; Pernyeszi et al., 2006; Zampori et al., 2009). Adsorption of the pollutants using organo-montmorillonite has attracted much attention as one of the removal methods.

Cationic surfactants intercalated montmorillonite have gradually been the subject of extensive research, and conventional cationic surfactants were widely used as modifiers, while Gemini surfactants which contain at least two hydrophilic and two hydrophobic groups were seldom used. In fact, the montmorillonite modified by Gemini surfactants was proved to be more efficiency in removing the organic pollutes (Rosen and Li, 2001; Li and Rosen, 2000) than the one modified by conventional cationic surfactant.

In this paper, montmorillonites modified by Gemini cationic surfactants which contain three hydrophobic and two hydrophilic groups, were prepared and characterized by FTIR, XRD and TG analysis. The *p*-chlorophenol adsorption test was also carried out, expecting to be useful in removing organic pollution in wastewater.

2. Materials and methods

2.1 Materials

The montmorillonite used in this study was from ZheJiang Huate Chemical Co. Ltd (Zhejiang, China). The cation exchange capacity (CEC) was 70 mmol/100 g. A new type of Gemini cationic surfactants in Fig.1, which were synthesized according to the literature (Chen et al., 2008a; Joanna and Jan., 2006), were used as modifiers.

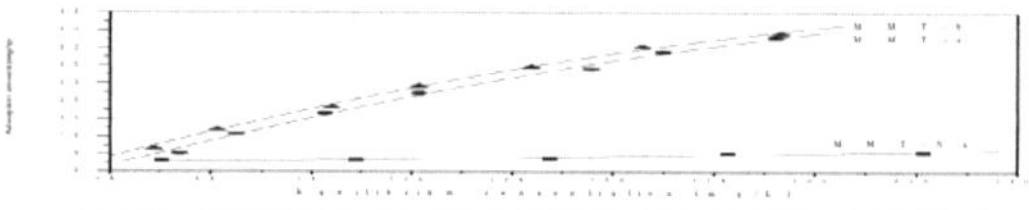

R1	n-$C_{11}H_{23}$	n-$C_{13}H_{27}$
R2	n-$C_{12}H_{25}$	n-$C_{14}H_{29}$
Surfactants	a	b

Fig.1. The structure of Gemini cationic surfactants

2.2 Preparation of modified montmorillonites

The modified montmorillonites were prepared as follows: Sodium montmorillonite power (5g) was dispersed in 100 mL of deionized water. After adjusting pH to 7.4 with hydrochloric acid, the dispersion was stirred at 80°C for 1 h. Gemini surfactants (2mmol) were dissolved in 50mL of deionized water and slowly dropped into the montmorillonite dispersion. After dropping, the dispersion was treated at 80°C for 4 h. The precipitates were separated by filtration and washed several times with deionized water until no chloride ion was detected by $AgNO_3$ solution. In this paper, MMT-a, b represent montmorillonite modified by surfactant a, b, respectively.

2.3 Characterization methods

2.3.1 Fourier transform infrared spectroscopy

The Fourier transform infrared (FTIR) spectra were obtained on FLA2000-104 FTIR spectrometer.

2.3.2 Thermogravimetric analysis

Thermogravimetric analysis (TG) were performed on a TGA Instruments (TGA /SDTA 85/e) and operated at a rate of 10.00°C/min from room temperature to 800.00°C in a high-purity flowing nitrogen atmosphere (100.0 mL/min).

2.3.3 X-ray diffraction

X-ray diffraction measurements (XRD) were performed using a D8-Advance X-ray diffractometer with CuKα radiation (λ = 1.5418 Å), operated at 40 KV and 30mA in a step scan mode. Diffraction measurements were conducted within the range of 2–10°, at a scanning speed rate of 0.02°/s.

2.4 Adsorption test

The adsorption experiments were carried out using batch equilibrium tests. 0.1g modified montmorillonite and 25 mL of prepared solution containing *p*-chlorophenol were placed in 250 mL conical flask. The adsorption experiments were carried out at

(35±0.2) °C and stirred at 180 r/min for 60 min. After separated by centrifugation, the concentration of *p*-chlorophenol was determined by Double Beam UV-VIS Spectrophotometers (λ_{max}= 280 nm).

3. Results and discussion

3.1 Characterization of modified montmorillonites

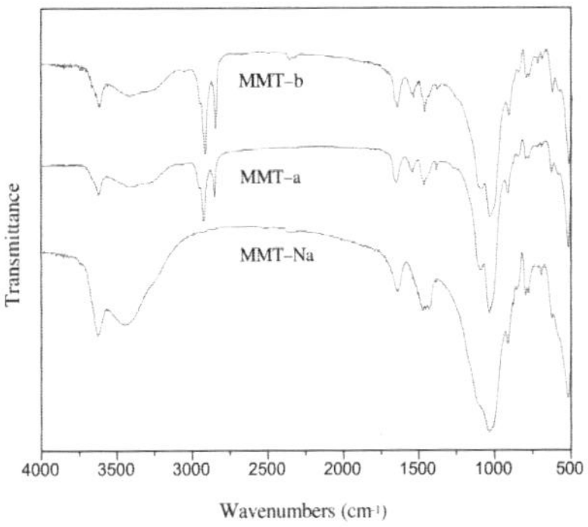

Fig.2. IR spectra of MMTs

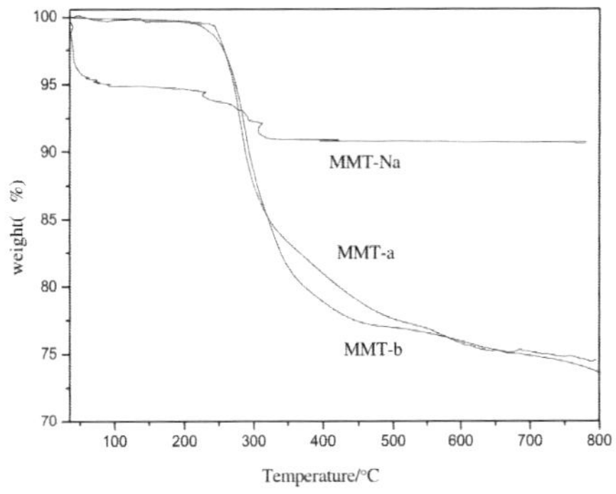

Fig.3. TG curve of MMTs

The unmodified montmorillonite is characterized by the bands of Si-O stretching and bending as well as OH bending (Zhou et al., 2009). The C-H stretching vibration of alkyl chains usually occur in the 3020-2800 cm^{-1} region (Kozak and Domka., 2004). In Fig.2, OH stretching vibration and bending vibration appears at 3640~3440 cm^{-1} region and 1639 cm^{-1} respectively. The absorption band at 1030 cm^{-1} is attributed to Si-O stretching vibration, while bands in 600~500 cm^{-1} are assigned to Si-O and Al-O bending vibration. The peaks at 2923 $cm^{-1}$and 2850 cm^{-1} are due to the asymmetrical and symmetrical C-H stretching of the methylene groups. The C-H bending vibration of the alkyl chains is observed at 1468 cm^{-1}. As is the Characteristic absorption peak of amide group, the absorption peaks that occur in 1652 cm^{-1} and 1545 cm^{-1} must be emphasized. Since there is no chloride ion to be detected, it can be proved that the surfactants have been successfully intercalated into the interlayer of montmorillonite.

Table 1. The weight loss in the thermogravimetric curves of MMTs

Samples	We.% loss at 35-200°C	We.% loss at 200-500°C	We.% loss at 750°C
MMT-Na	5.43	3.80	9.31
MMT-a	0.26	22.14	25.23
MMT-b	0.38	22.69	25.57

Thermo-analytical method provides a measure of the thermal stability of materials (Chen et al., 2008b) and has been widely used in studying the thermal decomposition of the modified montmorillonite(Xi et al., 2005a,b; Xie et al., 2001, 2002; He et al., 2005; Zhu et al., 2008). As shown by the TG curve in Fig.3, the thermal decomposition of MMT-Na occurs in two steps. The first step of weight loss is observed at 25-200°C region which is attributed to desorption of free water (water between particles and sorbed on the external surfaces of crystals) from the montmorillonite. The second step of weight loss occurs in 250-450°C regions, corresponding to the loss of water in the interlayer or to the dehydration of the hydrated cation in the interlayer. In Table 1, the organic constituent in modified montmorillonite mainly decompose at 200-500°C region, while there is little weight loss at 35-200°C region, which indicates that the free water has been replaced by Gemini cationic surfactants.

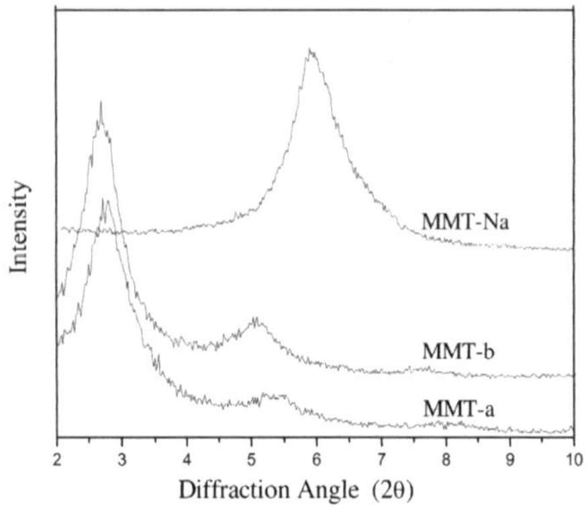

Fig.4. X-ray diffraction patterns of MMTs

According to Fig.4, the layer distance (Δd) was calculated by subtracting the sickness of the silicate layer of montmorillonite which equals 0.96 nm (Huskic et al., 2008), and listed in Table 2.

Table 2. The maximum diffraction angle and basal spacing of MMTs

Sample	Maximum diffraction angle 2θ/(°)	Basal spacing d001 (nm)	Layer distance Δd(nm)
MMT-Na	5.94	1.47	0.51
MMT-a	2.80	3.15	2.19
MMT-b	2.68	3.29	2.33

From Table 2, the basal spacing value of MMT-a and MMT-b is 3.15 nm and 3.29 nm respectively. As the carbon chain length of Gemini cationic surfactants increase, the size of the modifier augment, and the basal spacing also increases. Compared with the modifier Gemini cationic surfactant (12-4-12) reported by Zhou et al (2009), which increases the basal spacing to 2.06 nm, the modifiers used in this paper are more effective at expanding the interlayer space of the montmorillonite. According to literature (Lagaly, 1986), the alkyl chains in the modified montmorillonite interlayer might form monolayers(1.37nm), bilayers(1.77nm), pseudotrimolecular layers(2.17nm) and paraffin structure(2.77nm) as the basal spacing changes. We can conclude that these novel Gemini surfactants with more hydrophobic space groups are easier to form a paraffin structure in the montmorillonite interlayer than 12-4-12.

3.2 The adsorption isotherms of *p*-chlorophenol on montmorillonites

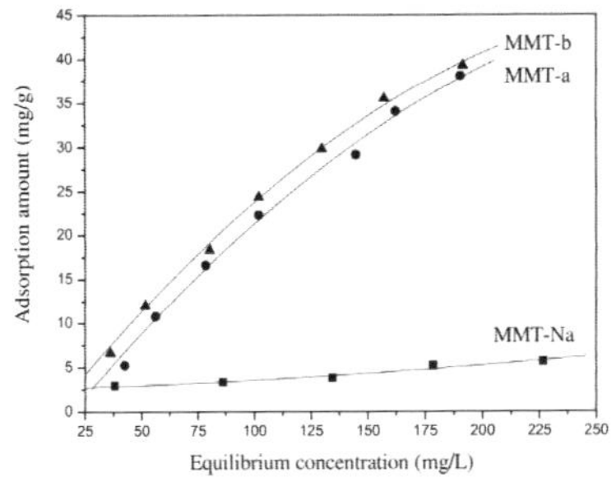

Fig.5. Isotherms of the adsorption of *p*-chlorophenol on MMTs

As shown in Fig 5, the adsorption amount of *p*-chlorophenol on MMT-Na has a linear relationship with equilibrium concentration, and the growth rate is extremely low. The modified montmorillonite did not have advantage in adsorption to *p*-chlorophenol at low equilibrium concentration, however, as the equilibrium concentration increase, the adsorption amount to *p*-chlorophenol of the modified montmorillonite increases rapidly. The adsorption amounts are up to 39.36 mg/g and 39.27 mg/g respectively at the maximum equilibrium concentration, which is seven times higher than the maximum adsorption amount of MMT-Na. It can be explained that there be more adsorption area and surface non-polarity of the montmorillonites modified by these novel Gemini surfactants. For the surface of MMT-Na is hydrophilic, it is prone to adsorb more water rather than *p*-chlorophenol, so little p-chlorophenol was adsorbed to the surface of MMT-Na, and the adsorption amount reveal slow growth as the equilibrium concentration increase. After modification, the surface of modified montmorillonite changes from hydrophilic to hydrophobic, the surface energy reduced. With a paraffin structure, the interlayer of the modified montmorillonite exhibited good compatibility with *p*-chlorophenol, adsorption and partition together affect the adsorption of *p*-chlorophenol on the modified montmorillonite, leading to high adsorption to *p*-chlorophenol.

4. Summary

FTIR analysis proves that the Gemini cationic surfactants have been successfully intercalated into the montmorillonite. The TG curve shows that there is little mass loss at 35-200°C regions, while the organic constitutes mainly decomposed in 200-500°C region. The XRD results show that the surfactants used in this paper can expand the basal spacing of the montmorillonite effectively and have more advantages than the other Gemini cationic surfactants. The two intercalators form a paraffin structure in the montmorillonite interlayer. The modified montmorillonite revealed a better removal of p-chlorophenol than the unmodified montmorillonite in adsorption test, which showed a promising application on the removal of the organic pollutes in wastewater.

References

Chen, F.S., Liu, X.M., Hu, Z.R., Zhao, C.X., 2008a. Synthesis and properties of cationic Gemini surfactants with amide group. *China Surfactant Detergent & Cosmetics.* 38 (5) 285-288.

Chen, D., Zhu, J.X., Yuan, P., Yang, S.J., Chen, T.H., He, H.P., 2008b. Preparation and characterization of anion-cation surfactants modified montmorillonite. *J. Therm. Anal. Cal.* 94, 841-848.

He, H.P., Ding, Z., Zhu, J.X., Yuan, P., Xi, Y.F, Yang, D., Frost, R.L., 2005. Thermal characterization of surfactant-modified montmorillonites. *Clays and Clay Minerals.* 53,287-293.

Huskic, M., Brnardic, I., Zigon, M., Ivankovic, M., 2008. Modification of montmorillonite by quaternary polyesters. *Journal of Non-crystalline Solids.* 354, 3326-3331

Joanna, W., Jan, C., 2006, Preparation, Surface-Active and Antielectrostatic Properties of Multiple Quaternary Ammonium Salts. *Journal of Surfactants and Detergents.* 9 (3), 221-226.

Juang, R-S., Lin, S-H., Tsao, K-H., 2004. Sorption of phenols from water in column systems using surfactant-modified montmorillonite. *Journal of Colloid and Interface Science.* 269, 46-52.

Kozak, M., Domka, L., 2004. Adsorption of the quaternary ammonium salts on montmorillonite. *Journal of Physics and Chemistry of solids.* 65, 441-445.

Lagaly, G., 1986. Interaction of alkylamines with different types of layered compounds. *Solid State Ionics.* 22, 43-51.

Lawrence, M.A.M., Kukkadapu, R.K., Boyd, S.A., 1998. Adsorption of phenol and chlorinated phenols from aqueous solution by tetramethylammonium- and etramethylphosphonium-exchanged montmorillonite. *Applied Clay Science.* 13, 13-20.

Li, F., Rosen, M.J., 2000. Adsorption of Gemini and Conventional Cationic Surfactants onto Montmorillonite and the Removal of Some Pollutants by the Clay. *Journal of Colloid and Interface Science.* 224, 265-271.

Pernyeszi, T., Kasteel, R., Witthuhn, B., Klahre, P., Vereecken, H., Klumpp, E., 2006. Organoclays for soil remediation: Adsorption of 2, 4-dichlorophenol on organoclay/aquifer material mixtures studied under static and flow conditions. *Applied Clay Science.* 32, 179-189.

Rosen, M.J., Li, F., 2001. The Adsorption of Gemini and Conventional Surfactants onto Some Soil Solids and the Removal of 2-Naphthol by the Soil Surfaces. *Journal of Colloid and Interface Science.* 234, 418-424.

Wibulswas, R., White, D.A., RAUTIU, R., 1999. Adsorption of phenolic compounds from water by surfactant-modified pillared clays. *Trans IChemE.* 77, 88-92.

Xie, W., Gao, Z.M., Pan, W.P., Hunter, D., Singh, A., Vaia, R., 2001. Thermal Degradation Chemistry of Alkyl Quaternary Ammonium Montmorillonite. *Chem.Mater.* 13, 2979-2990.

Xie, W., Xie, R.C., Pan, W.P., Hunter, D., Bryan,Koene., Tan, L.S., Vaia, R., 2002. Thermal stability of Quaternary phosphonium modified Montmorillonite. *Chem.Mater.* 14, 4837-4845.

Xi, Y., Martens, W., He, H., Frost, R.L., 2005a. Thermogravimetric analysis of organoclays intercalated with the surfactant octadecyltrimethylammonium bromide. *J. Therm. Anal. Cal.*

81,91-97.
Xi, Y.F., Frost, R.L., He, H.P., Kloprogge, T., Bostrom, T., 2005b. Modification of Wyoming Montmorillonite Surfaces Using a Cationic Surfactant. *Langmuir.* 21, 8675-8680.
Zampori, L., Gallo Stampino, P., Dotelli, G., 2009. Adsorption of nitrobenzene and orthochlorophenol on dimethyl ditallowyl montmorillonite: A microstructural and thermodynamic study. *Applied Clay Science.* 42, 605-610.
Zhou, L.M., Chen, H., Jiang, X.H., Lu, F., Zhou, Y.F., Yin, W.M., Ji, X.Y., 2009. Modification of montmorillonite surfaces using a novel class of cationic gemini surfactants. *Journal of Colloid and Interface Science.* 332, 16-21.
Zhu, R.L., Zhu, L.Z., Zhu, J.X., Xu, L.H., 2008. Structure of cetyltrimethylammonium intercalated hydrobiotite. *Applied Clay Science.* 42, 224-231.

(25a) Natural Gas Production From Gas Hydrates Coupled with Carbon Sequestration

Garrett C. Fitzgerald and Marco J Castaldi, Earth & Environmental Engineering, Columbia University, New York, NY

Natural Gas production via Gas Hydrates: A Parametric Study of Energy Efficiencies and Carbon Sequestration Potential.

Garrett Fitzgerald, Marco J. Castaldi

Abstract

While extensive fundamental research has been done, there has been little research focused on fuel production from hydrates coupled with CO_2 sequestration. A recently developed in-situ process for gas production from hydrates and COMSOL simulations show that it is possible to have an ideal energy efficiency of nearly 90% for land based reservoirs for hydrate loadings of 20%. Results from a 72 liter reactor vessel show production rates and efficiencies for hydrate loadings between 5 and 15 % pore volume conducted at both high [100 w] and low [20 w] relative heating rates. During the heating tests various trends were observed relating hydrate loading and heating rate to gas production rate and overall efficiencies. Results of multiple heating rates for in-situ dissociation tests also agree with published model simulations and indicate a low heat flux may preferred to yield the highest efficiency, however hydrate loading affects the optimal heating rate, this interaction is explored. We have now begun to quantify the energy associated with CO_2 hydrate formation and its ability to displace methane and remain sequestered. Recent findings of using CO_2 to displace methane in a hydrate reservoir during production of methane gas to be used as a fuel will be presented. Multiple CO_2 injection rates are investigated.

Introduction

Hydrates are non-stoichometric inclusion compounds which form ordered cages around light gas molecules that become entrapped in the lattice. These compounds are formed in the presence of water and guest molecules at sufficiently low temperature and high pressure. Natural gas hydrates are found in the form of methane hydrates around the globe in permafrost regions and in sediments below the deep-sea. The worldwide estimates of methane hydrates are massive and wildly uncertain. The USGS has estimated the in-place hydrates to be on the order of 400 million trillion cubic feet [TCF] , a staggering number compared to proven worldwide gas reserves of 5500 TCF.[**Demirbas et al.**] These large numbers are however only estimates of in-place hydrates and it is widely accepted that the recovery of CH_4 from hydrate formations is a technically challenging task and has yet to be proven economically feasible. Energy demand is expected to increase by 49% from 2007 to 2035 [IEO, 2010] and with this increase in consumption is an expected increase in CO_2 emission from 29 billion tons in 2007 to 44 billion tons in 2035. Methane hydrates present a unique opportunity to address both the energy consumption and CO_2 production dilemma that is inevitably in our near future.

CO_2 can be stored as hydrate during the simultaneous conversion of methane hydrates into methane gas for production. Due to the favorable thermodynamics of CO_2 hydrates this methods offers several benefits including the potential for a carbon neutral energy source as well as the potential for increased economic feasibility associated with a CO_2 assisted methane production. CO_2 injection also is advantageous to the continued geo-mechanical stability of the hydrate formation after CH_4 has extraction. Exploiting the fact that $CO\text{-}_2$ is more thermodynamically stable than CH_4 at the temperature and pressures present in natural hydrate deposits, not only will the CH_4 selectively dissociate, the CO_2 will replace the previous CH_4 hydrate re-stabilizing the sediment. The rates and extent to which this exchange process can occur is still unknown and is under current investigation and will be reported when data become available.

Experimental

There are currently three accepted methods for the recovery of CH_4 from hydrate deposits; thermal stimulation, formation depressurization and inhibitor injection. In this paper we investigate the overall energy efficiencies of several different thermal stimulation methods. Methane hydrate deposits vary widely in their defining characteristics such as sediment permeability, porosity, pore volume hydrate saturation, water saturation, and formation temperature and pressure. Hydrate saturation plays a key role in the rate of heat and mass transfer that can be achieved during the dissociation process. We have developed a large scale apparatus that is capable of dissociating hydrate via thermal stimulation, pressure reduction or a combination of the two at various hydrate saturations and under different heating conditions. The apparatus used in these experiments is depicted in figure1; this 70 liter internal volume setup is rated to 2000 [14 mpa] psi and is equipped with a centrally located resistive heater and CO_2 injection port. There are 12 k type thermocouples arranged to capture heat flow patterns in the radial and azimuthal directions. The sediment used in these experiments is 500 micron quartz sand that is probed with two gas sampling ports for gas composition analysis.

Hydrate formation occurs in a stochastic and heterogeneous manner and thus the precise location of hydrate in the sediment is somewhat unknown. This uncertainty requires that great care be taken when forming the hydrate as well as during data analysis to insure no faulty conclusions are made. When growing the hydrates in the lab we chose to use a gas injection method. The sediment is partially saturated with water and brought to the desired hydrate formation temperature prior to injection. When the entire system has reached thermal equilibrium CH_4 is injected from the bottom of the vessel and allowed to percolate up through the wet sediment and eventually fill the 2 liter headspace above the sediment. When the desired overpressure above the hydrate stability zone (HSZ) is reached, gas flow is terminated and the hydrate begins to form. The process is repeated when the pressure stabilizes indicating hydrate-gas-water equilibrium; the number of cycles is increased to achieve higher hydrate saturations.

When the hydrate is formed with known hydrate, water, and gas saturations the heating test are initiated. The heat flux in is set constant at either 20 [low] watts or 100[high] watts using a custom omega resistive heating element. The thermal stimulation causes the hydrate to dissociate into the gas phase resulting in a pressure rise and gas accumulation in the head space. The pressure is allowed to rise ~30 psi [200 kpa] at which point a back pressure regulator releases the gas until the pre-heating pressure is realized. Mass flow meters and the Peng-Robinson EOS are used to complete a mass balance and determine the amount of hydrate that dissociated from the thermal stimulation.

Results

The preliminary results from the heating test show that at higher hydrate saturations higher heat flux is more efficient; as well it was observed that the efficiency for the lower heat flux is greater at for the lower hydrate saturation. These numbers are preliminary and will be verified in a second round of tests to assure accuracy. Due to the low thermal conductivity of pure hydrate the overall conductivity of the formation decreases for increased hydrate loadings, this leads to a buildup of heat close to the heater which raises the local formation temperature significantly above the dissociation temperature. This excessive heating of the non gas producing sediment is one source of the energy losses and decreased efficiencies associated with the increased hydrate saturation; it is essentially a competition between the dissociation of the hydrate and the thermal diffusivity of the sediment.

Using an high heat flux can also suffer from the same issue of overheating the non-productive matrix, however at the high heating rate this was not observed for the higher hydrate loading because the increased flux was spent in dissociating the greater volume of hydrate present. Test 4 was compromised and must be repeated. The full set of data will be presented and discussed in detail.

Test 1 Q=100 15.6 % Hydrate eta = 69%	Test 2 Q=100 10.2% Hydrate eta=?
Test 3 Q=20 15.6 % Hydrate Eta=49%	Test 4 Q=20 10% Hydrate Eta = 71%

Results from in-situ heating and CO_2 sequestration will also be presented; similar tests to those described above will be reported for various heating rates and various CO_2 injection rates.

Conclusion

CH_4 production via the thermal stimulation method was conducted using a resistive heating element in hydrate bearing sediment of different saturations. Hydrate saturation and heating rate have important impacts on the efficiency of production. Increased heating with higher hydrate loading was shown to be more efficient that lower heating at the same saturation. However the is competition between hydrate loading and heating rate and the heating rate must be designed to match the thermal diffusion rate in the hydrate bearing sediment.

References:

Demirbas, Ayhan, 'Methane hydrates as potential energy resource: Part 1 – Importance, resource and recovery' Energy Conversion and Management, Vol 51, 7 2010

IEO, 2010 International Energy Outlook 2010.< www.eia.gov/oiaf/ieo/index.html.> Energy Information Administration 2010.

Boswell, Ray. 'Resource potential of methane hydrate coming into focus' Original Research Article *Journal of Petroleum Science and Engineering, Volume 56, Issues 1-3, March 2007, Pages 9-13*

See more of this Session: Greenhouse Gases Capture and Sequestration

See more of this Group/Topical: Topical 6: 11th Topical Conference on Gas Utilization

(25b) Molecular Modeling of Structure and Transport of CO_2 In Organic-Based Systems

Mahnaz Firouzi and Jennifer Wilcox, Department of Energy Resources Engineering, Stanford University, Stanford, CA

CO_2 capture and sequestration is considered a primary, large-scale option for atmospheric CO_2 concentration stabilization. The aim of the current work is to fundamentally understand the molecular processes relevant to enhanced natural gas recovery from coal and gas shales using CO_2 with potential carbon storage. Molecular simulations have been carried out to model the organic matrix comprising coal and gas shales' microfracture and related pore networks that serve as mechanisms of gas transport. To study these phenomena, molecular dynamics simulations have been carried out to investigate the transport properties of pure carbon dioxide, methane and nitrogen as well as binary mixtures consisting of nitrogen and carbon dioxide and also methane and carbon dioxide. The carbon models are represented by a three-dimensional pore space, generated atomistically by Voronoi tessellation of the space, using tens of thousands of atoms. We use non-equilibrium molecular dynamics simulations with an external chemical potential or pressure gradient imposed on the system. Extensive simulations are carried out to study the effect of the pore structure including defected graphite and also microfractures. The permeability results are compared with available experimental data. These fundamental studies will lead to an understanding of the mechanism associated gas transport in these confined carbon spaces and could potentially lead to the enhanced recovery of natural gas with the potential for CO_2 storage in coalbeds and gas shales.

See more of this Session: Greenhouse Gases Capture and Sequestration
See more of this Group/Topical: Topical 6: 11th Topical Conference on Gas Utilization

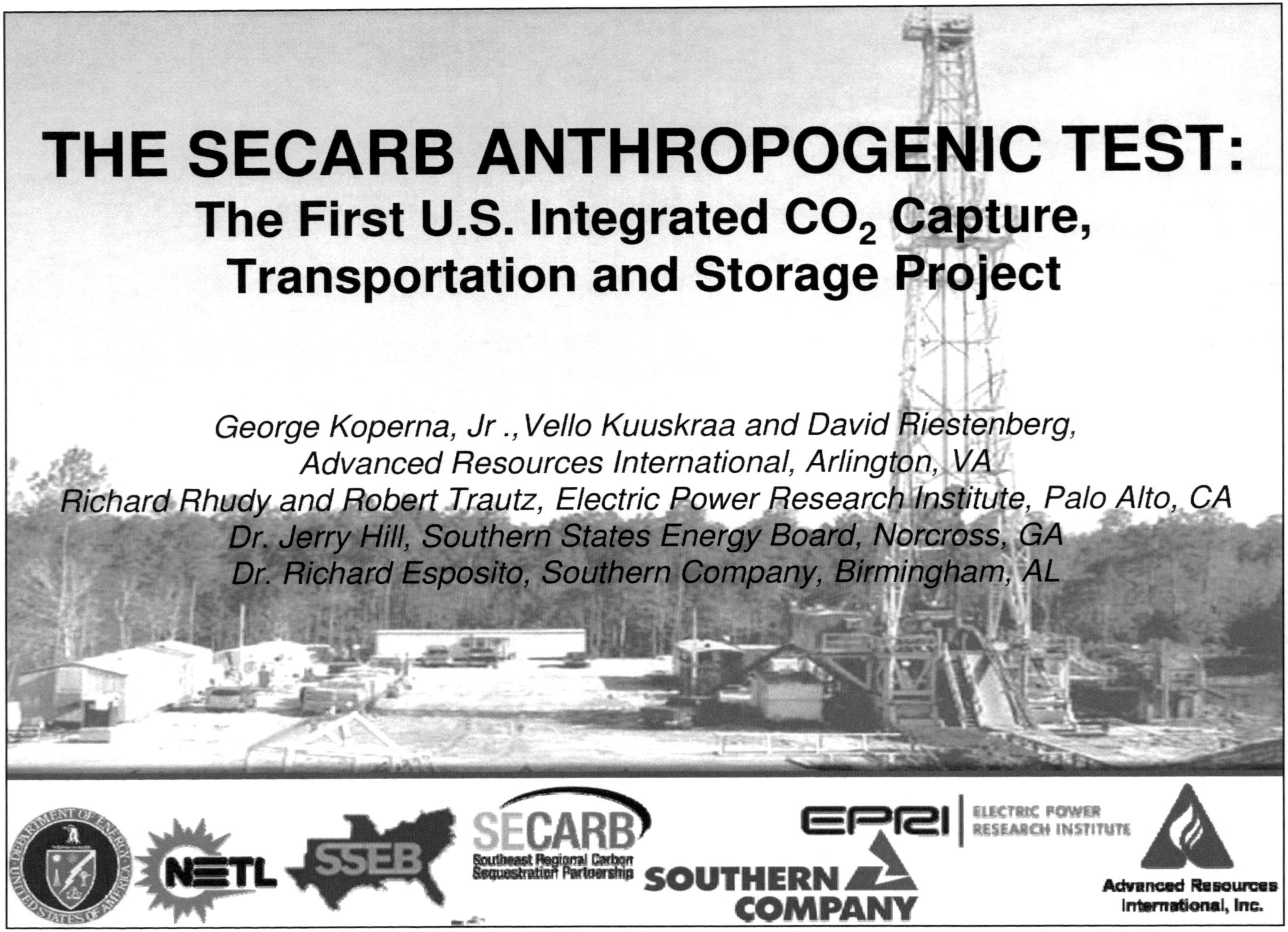
THE SECARB ANTHROPOGENIC TEST:
The First U.S. Integrated CO2 Capture, Transportation and Storage Project
George Koperna, Jr ., Vello Kuuskraa and David Riestenberg,
Advanced Resources International, Arlington, VA
Richard Rhudy and Robert Trautz, Electric Power Research Institute, Palo Alto, CA
Dr. Jerry Hill, Southern States Energy Board, Norcross, GA
Dr. Richard Esposito, Southern Company, Birmingham, AL
NETL
SSEB
SECARB
Southeast Regional Carbon Sequestration Partnership
EPRI
ELECTRIC POWER RESEARCH INSTITUTE
SOUTHERN COMPANY
Advanced Resources International, Inc.

Acknowledgement

This presentation is based upon work supported by the Department of Energy National Energy Technology Laboratory under **DE-FC26-04NT42590**. This report was prepared as an account of work sponsored by an agency of the United States Government. Neither the United States Government nor any agency thereof, nor any of their employees, makes any warranty, express or implied, or assumes any legal liability or responsibility for the accuracy, completeness, or usefulness of any information, apparatus, product, or process disclosed, or represents that its use would not infringe privately owned rights. Reference herein to any specific commercial product, process, or service by trade name, trademark, manufacturer, or otherwise does not necessarily constitute or imply its endorsement, recommendation, or favoring by the United States Government or any agency thereof. The views and opinions of authors expressed herein do not necessarily state or reflect those of the United States Government or any agency thereof.

Presentation Outline

1. DOE's Regional Carbon Sequestration Partnerships
2. "Lessons Learned" from SECARB's Phase II Activities
3. SECARB's Phase III Anthropogenic Test
4. Selecting a Safe, Secure CO_2 Storage Site
5. Detailed Reservoir Characterization
6. Reservoir Modeling
7. Making Major Investments in CO_2 Monitoring
8. Addressing Risks and Mitigation Plans
9. CO_2 Capture Facility

11AIChE_Koperna

Regional Carbon Sequestration Partnerships

"Developing the Infrastructure for Wide Scale Deployment"

7 Partnerships, 500+ distinct organizations, 43 States, 4 Canadian Provinces

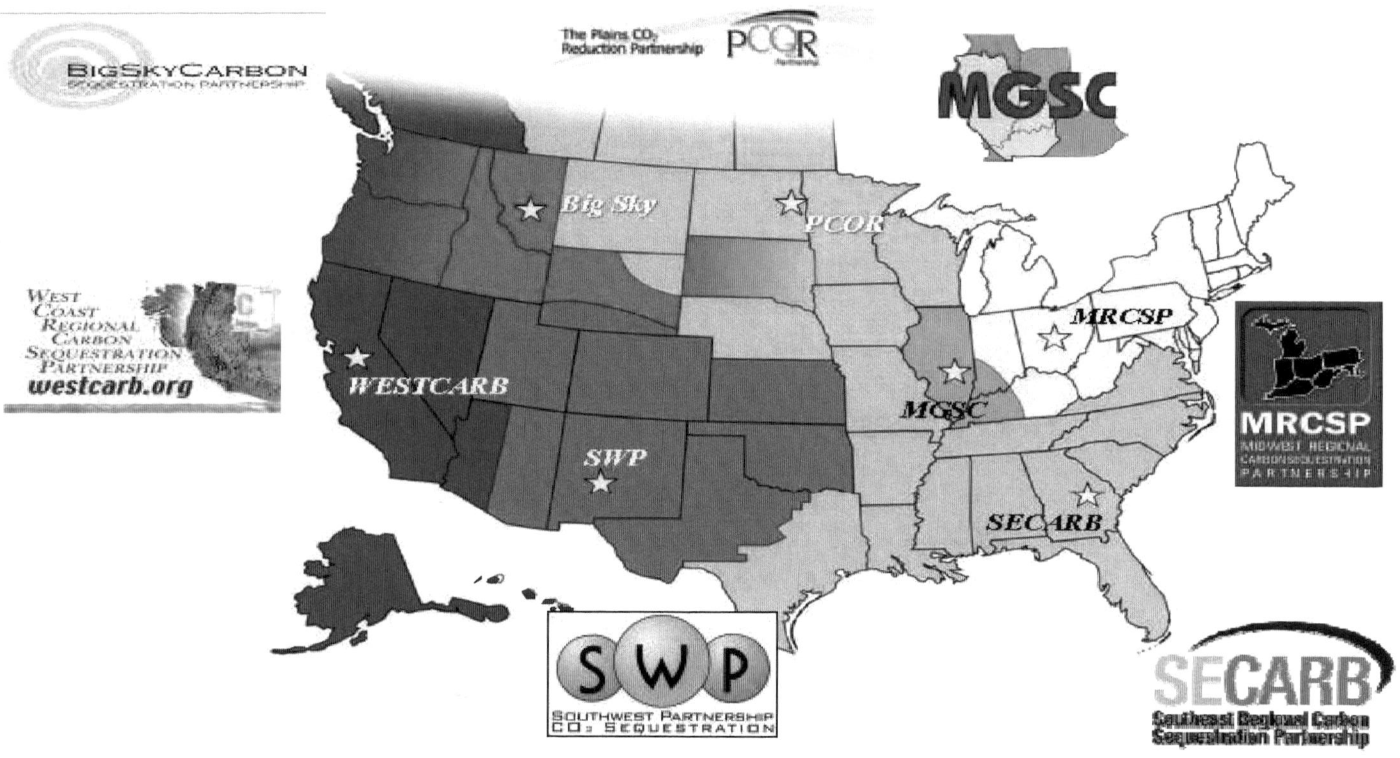

11AIChE_Koperna

Building on a Foundation of Success

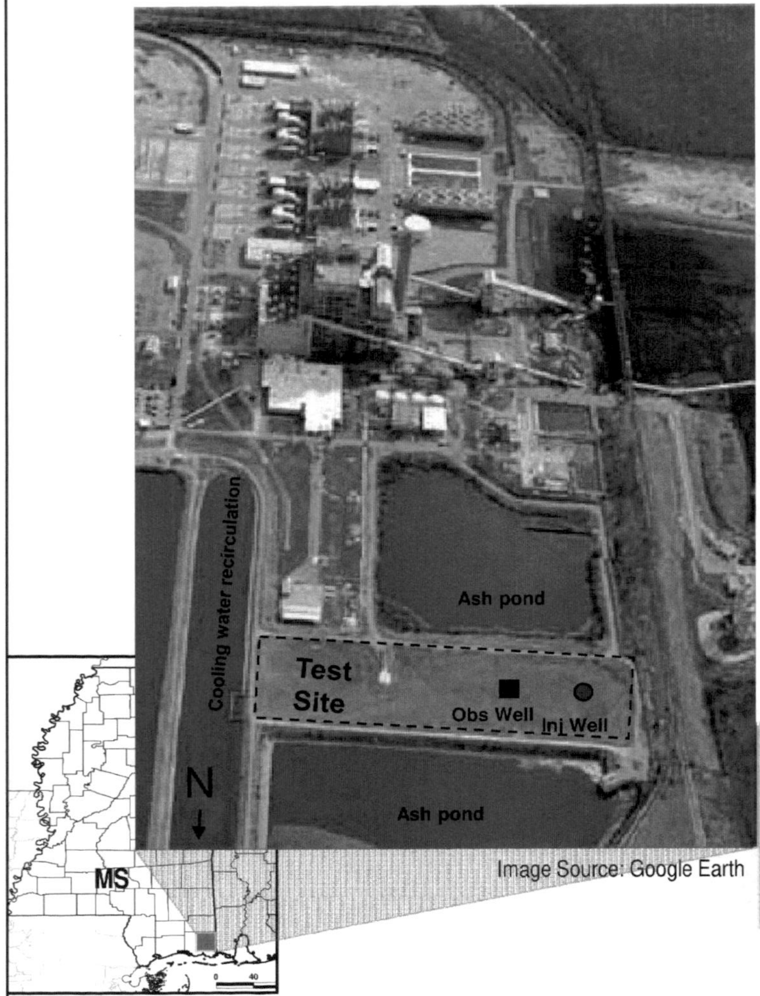

Image Source: Google Earth

SECARB's Phase II Mississippi Test Site
(Mississippi Power's Plant Daniel)

- Established viability of regionally significant and areally extensive Tuscaloosa Massive Sand for CO_2 injection and storage.
- Built experience and confidence for operating CO_2 storage at an operating power plant site.
  - Drilled and characterized two deep wells
  - Injected 3,027 tons of CO_2 at 180 tons per day (3 MMcfd), trucked to site.

11AIChE_Koperna

SECARB Phase III Demonstrations

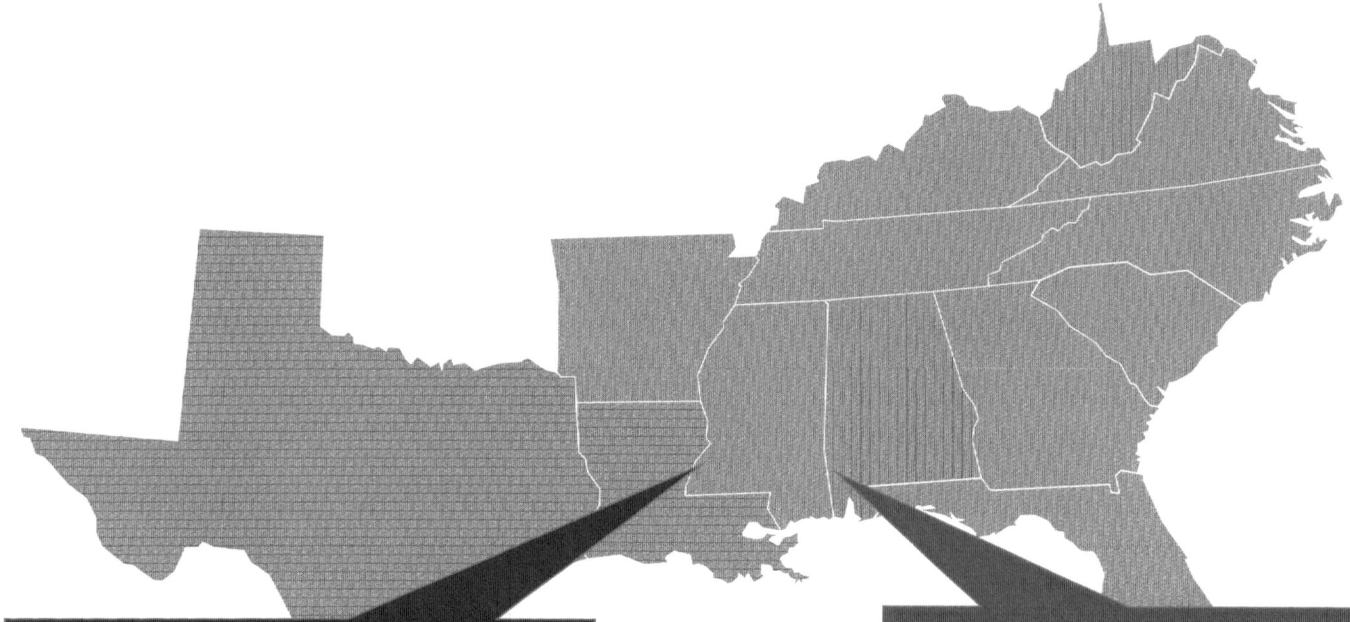

Early Test

Denbury Resources' Cranfield Field
Near Natchez, Mississippi

Gulf Coast Carbon Center

Anthropogenic Test

Capture: Alabama Power Plant Barry, Bucks, Alabama

Transportation: Denbury Onshore, LLC

Geo Storage: Citronelle Dome, Citronelle, Alabama

EPRI ELECTRIC POWER RESEARCH INSTITUTE

11AIChE_Koperna

The Anthropogenic Test Is Pursuing Ambitious Goals and Objectives

In support of SECARB's overall goal - - test and demonstrate safe, secure CO_2 injection and storage in regionally significant saline reservoirs - - the Anthropogenic Test has four objectives:

1. Conduct a fully integrated CCS project (capture, transport and storage)
2. Test the CO_2 flow, trapping and storage mechanisms of the Paluxy Fm., a regionally extensive Gulf Coast saline formation.
3. Demonstrate how a saline reservoir's architecture can be used to maximizing CO_2 storage and minimize the areal extent of the CO_2 plume.
4. Test the adaptation of commercially available oil field tools and techniques for monitoring CO_2 storage (e.g., VSP, cross-well seismic, cased-hole neutron logs, tracers, pressure, etc.)

11AIChE_Koperna

The Anthropogenic Test Schedule

- Well drilling operations began in late December 2011
- Baseline monitoring activities will occur in the spring of 2011
- CO_2 injection operations begin in the third quarter of 2011, continue for 2 to 3 years
- 3 years of post-injection monitoring

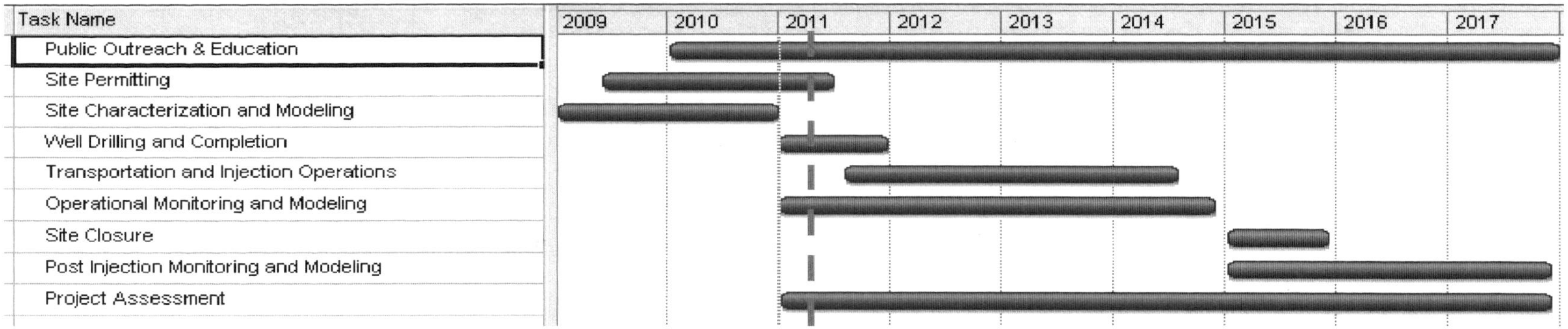

11AIChE_Koperna

SECARB's Phase III Anthropogenic Test

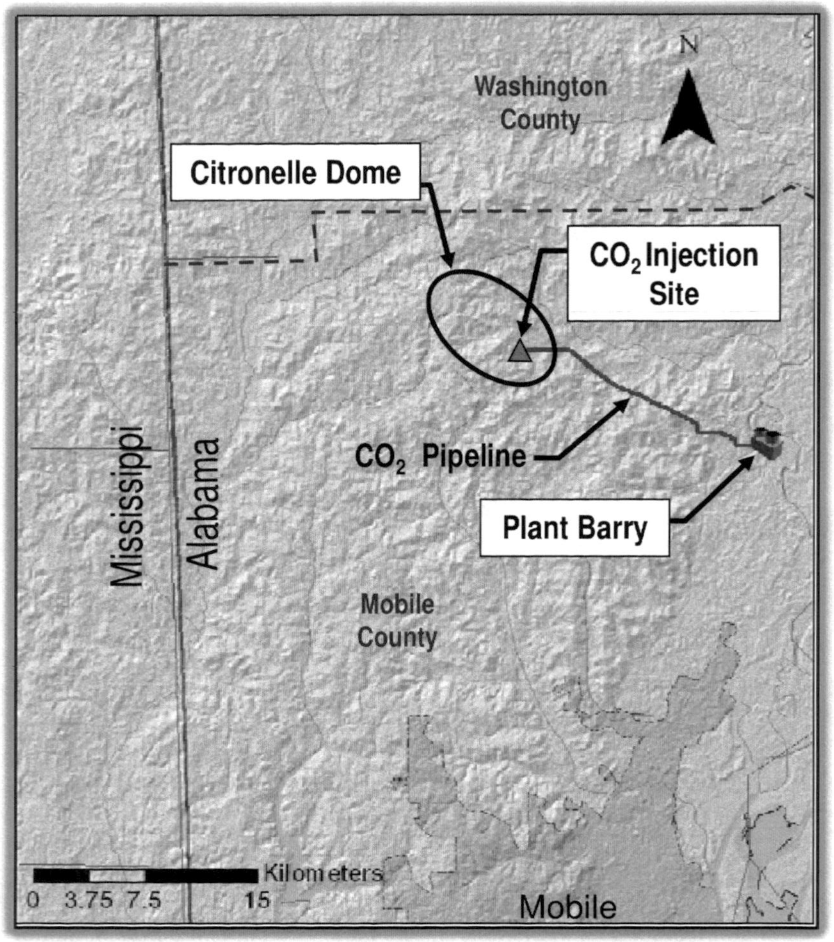

A Fully Integrated CCS Project

The 25 MW CO_2 capture unit at Alabama Power's (Southern Co.) Plant Barry will become operational in 3Q 2011.

A newly built 12 mile CO_2 pipeline will transport CO_2 from Plant Barry to the Citronelle Dome.

From 100 to 300 thousand metric tons of CO_2 will be injected into the Paluxy saline formation over 2 to 3 years.

Advanced Resources and supporting researchers will conduct 3 years of monitoring after CO_2 injection and then close the site.

11AIChE_Koperna

Selecting a Safe, Secure CO_2 Capacity Site

Citronelle Dome
Saline Reservoirs and Seals

System	Series	Stratigraphic Unit	Major Sub Units	Potential Reservoirs and Confining Zones
Tertiary	Plio-Pliocene		Citronelle Formation	Freshwater Aquifer
Tertiary	Miocene	Undifferentiated		Freshwater Aquifer
Tertiary	Oligocene		Chicasawhay Fm.	Base of USDW
Tertiary	Oligocene	Vicksburg Group	Bucatunna Clay	Local Confining Unit
Tertiary	Eocene	Jackson Group		Minor Saline Reservoir
Tertiary	Eocene	Claiborne Group	Talahatta Fm.	Saline Reservoir
Tertiary	Eocene	Wilcox Group	Hatchetigbee Sand Bashi Marl Salt Mountain LS	Saline Reservoir
Tertiary	Paleocene	Midway Group	Porters Creek Clay	Confining Unit
Cretaceous	Upper	Selma Group		Confining Unit
Cretaceous	Upper	Eutaw Formation		Minor Saline Reservoir
Cretaceous	Upper	Tuscaloosa Group	Upper Tusc. / Mid. Tusc.	Minor Saline Reservoir
Cretaceous	Upper	Tuscaloosa Group	Marine Shale	Confining Unit
Cretaceous	Upper	Tuscaloosa Group	Lower Tusc.: Pilot Sand Massive sand	Saline Reservoir
Cretaceous	Lower	Washita-Fredericksburg	Dantzler sand	Saline Reservoir
Cretaceous	Lower	Washita-Fredericksburg	Basal Shale	Primary Confining Unit
Cretaceous	Lower	Paluxy Formation	'Upper' 'Middle' 'Lower'	Proposed Injection Zone
Cretaceous	Lower	Mooringsport Formation		Confining Unit
Cretaceous	Lower	Ferry Lake Anhydrite		Confining Unit
Cretaceous	Lower	Donovan Sand	Rodessa Fm. 'Upper'	Oil Reservoir
Cretaceous	Lower	Donovan Sand	'Middle'	Minor Saline Reservoir
Cretaceous	Lower	Donovan Sand	'Lower'	Oil Reservoir

The Anthropogenic Test's CO_2 Storage Site

The saline formation selected for CO_2 injection and storage is regionally extensive, but geologically challenging.

- Proven four-way closure at Citronelle Dome.
- Deep Lower Cretaceous Paluxy Fm (at 9,400').
- Massive 1,100 foot interval of stacked fluvial sands and shales.
- Numerous overlying confining units (at least 5).
- No evidence of faulting or fracturing, based on reinterpretation of existing 2D seismic lines.

However, prior to drilling the characterization well, only limited data exist on the sand continuity, porosity and permeability of the Paluxy Fm at Citronelle Dome.

11AIChE_Koperna

Detailed Reservoir Characterization

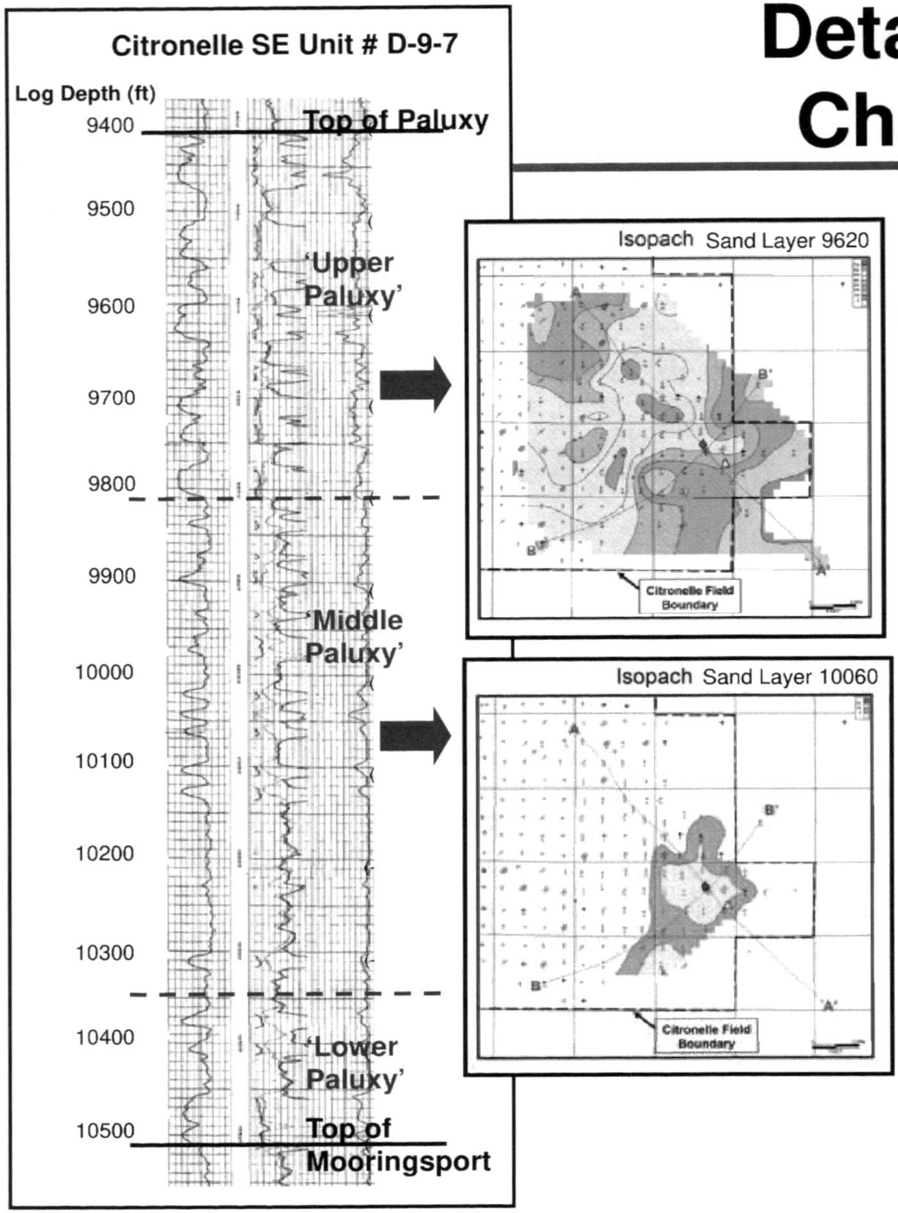

Establishing Reservoir Properties for Paluxy Saline Formation

Given the challenges posed by a thick, fluvial sequence of sands and shales, the project invested considerable time and effort on reservoir characterization:

- Sand continuity mapping to determine "open" or "closed" sand intervals.
- Detailed analysis of over 80 well logs for porosity and depositional style.
- Cross-plots of porosity versus permeability.

We also collected and analyzed core and log data on the Paluxy Fm in other parts of the Gulf Coast region.

11AIChE_Koperna

Detailed Reservoir Characterization (Cont'd)

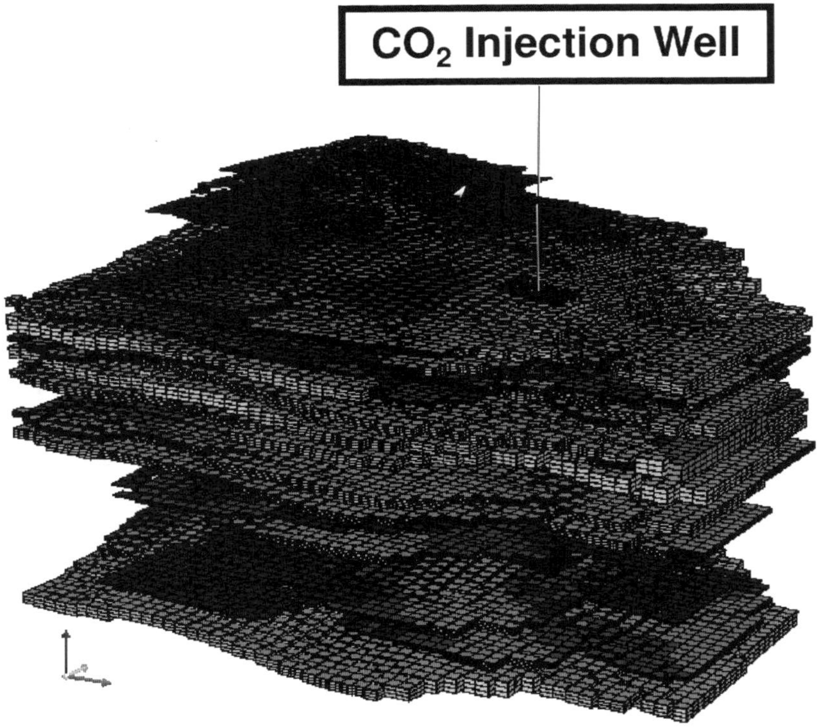

Building the Geologic Model

Based on detailed characterization of the thick Paluxy sand/shale interval, we selected 17 sand units for CO_2 injection:

- 343 net feet of sand
- Average sand porosity of 20%
- Permeability from 50 to 75 millidarcies
- Normal pressure and temperature gradients

Subsequent reservoir characterization (currently underway) will help confirm and update the key CO_2 injectivity and storage capacity parameters.

11AIChE_Koperna

Detailed Reservoir Characterization (Cont'd)

Confirming and Refining Our Reservoir Characterization

At the end of December 2010, we began drilling the Paluxy reservoir characterization well, D9-8 #2, at the Citronelle Dome.

- Well drilled to 11,800' TD (spud to TD) in 30 days and under budget.
- Recovered extensive whole core (98 feet in two intervals) plus 45 sidewell cores in the two confining units, the above primary seal zone and the remaining CO_2 storage zone.
- Ran full set of logs (Triple Combo, MRI, Mineralogy, Dipole Sonic, CBL, etc.).

Well was cased and cemented in late January, 2011 and will be used as an observation/monitoring well

11AIChE_Koperna

Reservoir Modeling

Modeling the CO_2 Plume

X-Sectional View of CO_2 Plume CO2 Injection Well

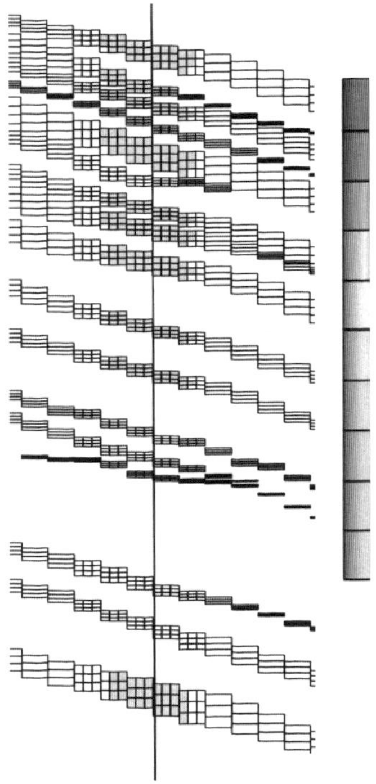

3D View of CO_2 Plume CO2 Injection Well

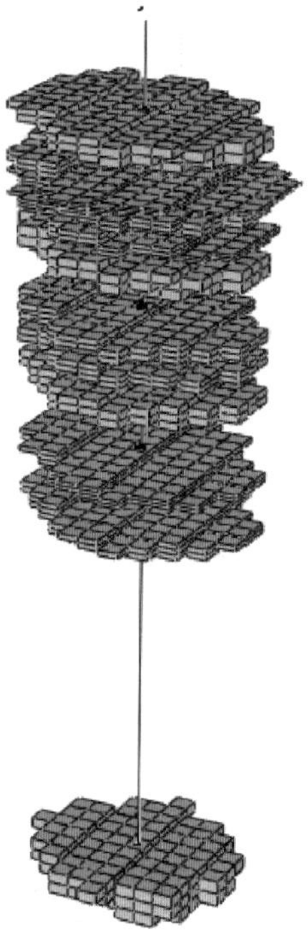

Understanding CO_2 Flow and Optimizing Storage Capacity

The information from detailed reservoir characterization is being used to model and optimize the CO_2 plume:

- Areal extent of CO_2 (less than 1,000' in any direction) will be limited by injection into multiple sand layers (taking advantage of reservoir architecture).
- Low formation dip and relatively uniform reservoir properties imply a near-circular plume and little post CO_2 injection up-dip migration.

We will conduct a series of additional reservoir simulations to calibrate the geologic model and support MVA as the project progresses.

11AIChE_Koperna

Making Major Investments in CO_2 Monitoring

CO_2 Injection and Storage Site

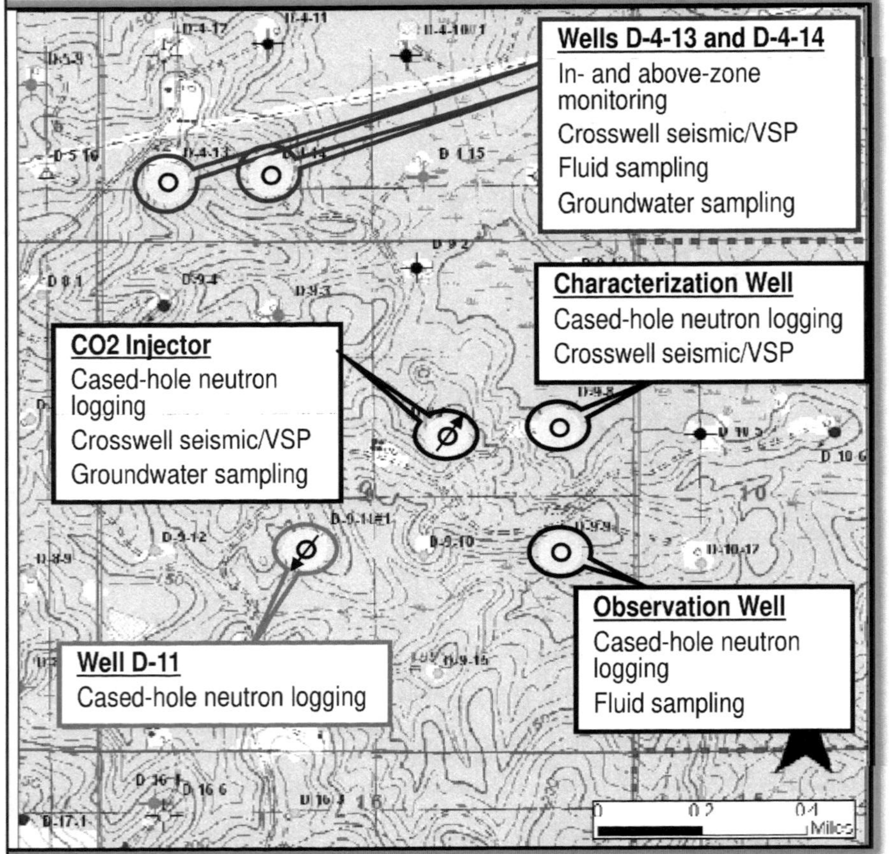

CO_2 Monitoring Design

The anthropogenic test will use five deep plus three shallow monitoring wells for tracking the CO_2 plume:

- Near-surface and deep reservoir fluid sampling.
- In-zone and above-zone pressure and temperature monitoring.
- Cased-hole neutron logging.
- Crosswell seismic and VSP.

We are indebted to SECARB's Early Test/BEG for monitoring guidance and to Denbury Onshore LLC for contributing three of their existing wellbores for monitoring.

11AIChE_Koperna

Addressing Risks and Mitigation Plans

The Anthropogenic Test has spent considerable effort to address risks and incorporate consideration of risks into the project plan, including:

1. ***Detailed site selection and characterization***
   - Four-way closure at Citronelle Dome.
   - No evidence of faults or major fractures (based on existing seismic).
   - Multiple (5) confining units and seals.
   - Thick package of sands and shales.
2. ***Rigorous reservoir modeling of the CO_2 plume***
   - Incorporation of reservoir architecture to limit areal extent of CO_2 plume.
   - Understanding of CO_2 trapping and immobilization mechanisms.
   - Plans for calibrating the reservoir model with monitoring data.

11AIChE_Koperna

Addressing Risks and Mitigation Plans (Cont'd)

3. ***Selecting high quality well drilling and service vendors***
   - Used Denbury Onshore LLC, experienced in drilling/completing CO_2 injection wells.
   - Used high quality well casing (L-80), two stage cementing and CO_2 resistant cement.
   - Integrated experienced ARI staff and site construction design with "first rate" service vendors.

4. ***Comprehensive program of monitoring***
   - Multiple cross-well seismic and consideration of VSP.
   - Adaptation of oil field tools to monitor CO_2 (e.g., cased-hole neutron log, fluid sampling, downhole P/T).
   - Collaboration with CO_2 Capture Project for testing innovative MVA technology.

We participated with DNV on incorporating our test site procedures into the IEA "risk registry" and are documenting our risk mitigation procedures for site construction, operations and closure.

11AIChE_Koperna

Capture Unit at Alabama Power's Plant Barry

MHI advanced amine capture unit

- 25 MW post combustion slip stream
- Fabricate off-site and barge to Plant Barry
- Compress CO_2 to 2000 psi
- Scheduled start up during summer, 2011
- Separately funded

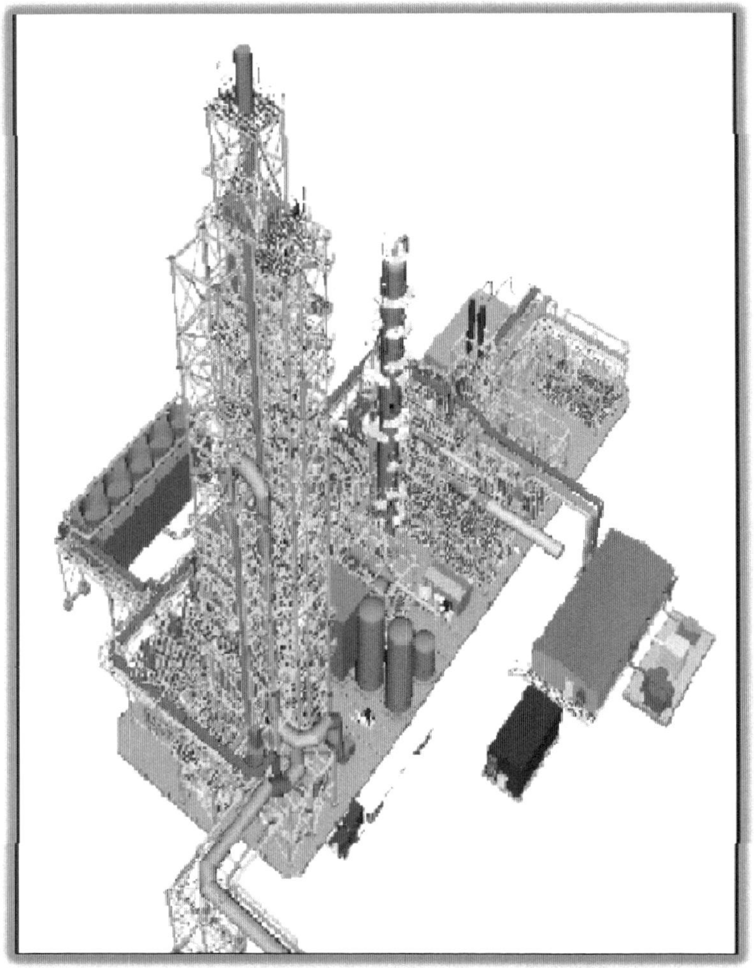

11AIChE_Koperna

Simplified CO_2 Scrubbing Process (Amine)

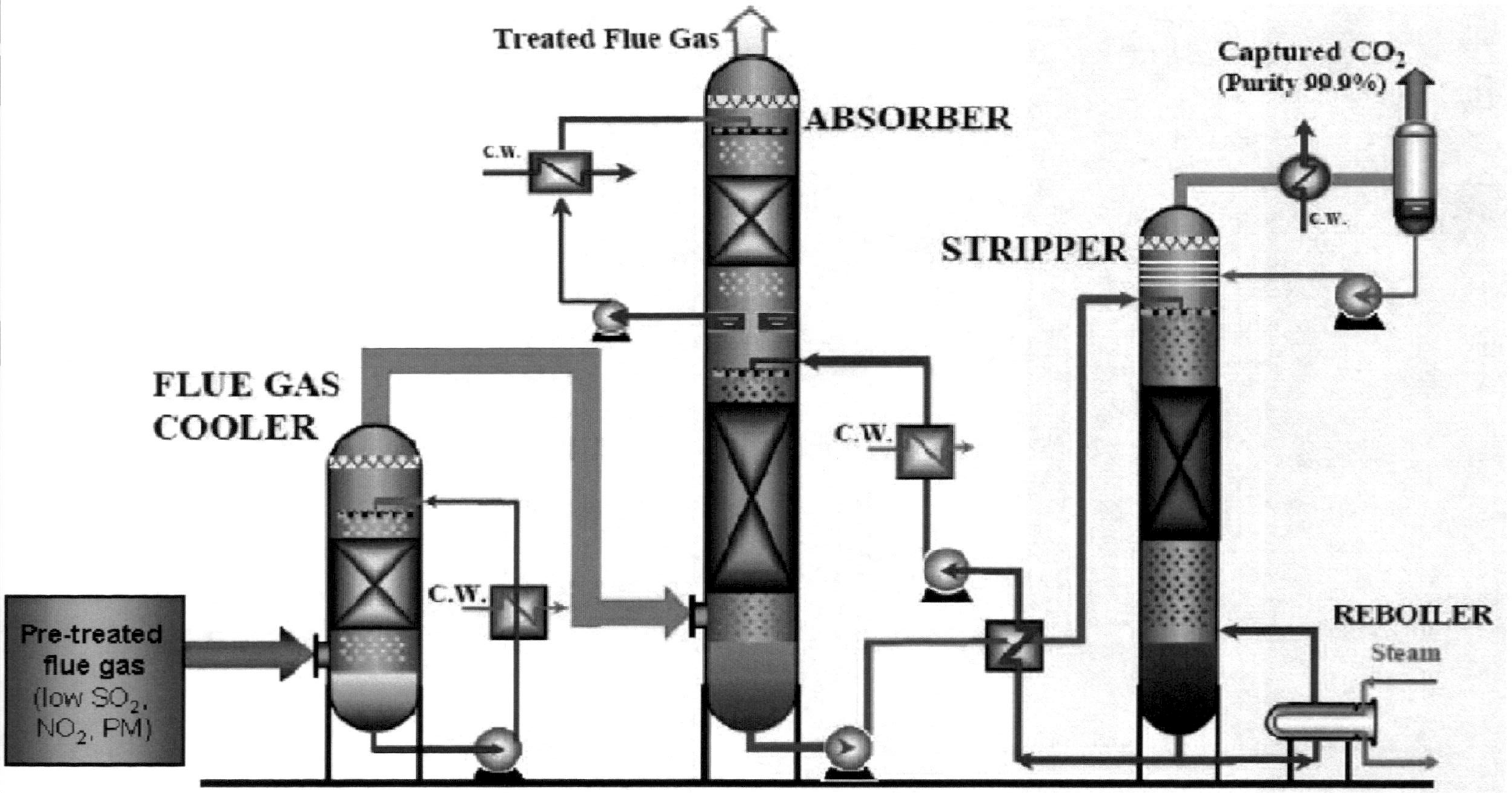

Key Points: needs >99% SO_2 removal; consumes process steam

11AIChE_Koperna

19

Groundbreaking Ceremony: Capture Unit

Alabama Power's Plant Barry, April 2010, Bucks, Alabama

SOUTHERN COMPANY

MITSUBISHI HEAVY INDUSTRIES, LTD.

EPRI | ELECTRIC POWER RESEARCH INSTITUTE

11AIChE_Koperna

Unit Transportation (9-2010)

11AIChE_Koperna

Site Progress Photos (11-2010)

11AIChE_Koperna

(25d) Effect of Flue Gas Contaminants on CO_2 Sequestration In Saline Formation

Yee Soong[1], S. W. Hedges[1], B. H. Howard[1], J. Baltrus[1], C. A. Summers[1], W. K. O'Conner[1] and D. Allen[2], (1)National Energy Technology Laboratory, U.S. Department of Energy, Pittsburgh, PA, (2) Geological Sciences, Salem State College, Salem, MA

A great deal of concern has been expressed with regard to global climate change and its link to growing atmospheric concentrations of carbon dioxide. To decrease the impact of anthropogenic CO_2 on global climate, several strategies are under development that will potentially remove CO_2 from the atmosphere or decrease CO_2 emission. One such strategy involves the capture of CO_2 from large point sources (such as fossil fuel-fired power plants) and the long-term storage of CO_2 underground. Carbon dioxide may be sequestered in various geological formations, including depleted oil and gas reservoirs, unmineable coal seams, basalt formations, and deep saline aquifers. Deep saline aquifers have been reported to have the most estimated capacity for CO_2 sequestration. Most currently sequestration studies in saline formation are focused on pure CO_2. However, the impacts of contaminants such as O_2, SOx and NOx, typical of flue gas streams in post-composition systems, are often neglected. This study examines the impact of flue gas contaminants on saline formation under CO_2 sequestration conditions.

Most CO_2 captured from power plants, especially coal-fueled power plants, will inevitably contain some level of various contaminant species that will react chemically differently than pure CO_2 in saline formation. For example, if SO_2 is present in the gas stream, sulfurous/sulfuric acid will from in the presence of water in the reservoir. This will further reduce the pH of the brine in the system. Mineral species in the formation rock, such as K-feldspar, calcite, kaolinite, etc., that are stable in the native environment but not stable at low pH will dissolve, and new product mineral phases possibly precipitate as the pH drops in the system. Therefore, there is a need to investigate the impact of flue gas contaminants on saline formation under CO_2 sequestration conditions.

In this study, both simulation and experimental works were conducted. The simulation work was conducted via use of Geochemist workbench. We simulated the chemical reactions in the Mt. Simon sandstone environment with exposure to CO_2 mixed with other gas species under conditions that mimic sequestration. (Experimental conditions were as follows: solid material - Mt. Simon sandstone; liquid - Illinois Basin brine; T and P - 50 C, 1500 psi; gas composition - 1% SO_2, 4% O_2, 95% CO_2). The simulated experimental conditions include various Mt. Simon/Ill brine ratios from 3, 5 and 15 in a one liter vessel both with and without SO_2 and O_2. Preliminary results showed that, as expected, the addition of SO_2 and O_2 affected pH, the concentrations of cations in the aqueous phase, and the oxidation state of iron. Fluid pH was acidic in all simulations, although though it was markedly lower in the simulations with SO_2 and O_2. Increasing the proportion of solid material in the system buffered the system and increased pH. Ca and Mg ions responded differently to the addition of SO_2 and O_2 and showed a strong dependence on pH in the mixed gas system. Aqueous ferrous iron and siderite were predicted by the CO_2-alone simulations.

Selected conditions were examined experimentally in an autoclave reactor. Reactions were conducted in a closed, well mixed reactor with CO_2 or $CO_2/O_2/SO_2$ flue gas product mix at the identical conditions (50 °C, 1,500 psi, 300 R.P.M. and 7 days of reaction time). The results were compared with the simulation predictions.

See more of this Session: Greenhouse Gases Capture and Sequestration
See more of this Group/Topical: Topical 6: 11th Topical Conference on Gas Utilization

(25e) CO_2 Capture From Binary Gas Mixtures Using Elastic Layered MOF Adsorbents

Christian M. Lastoskie, Civil & Environmental Engineering, University of Michigan, Ann Arbor, MI and Tran D. Trinh, Civil and Environmental Engineering, University of Michigan, Ann Arbor, MI

Elastic layered metal-organic framework (ELM) adsorbents exhibit a flexible two-dimensional latent porous crystalline structure. ELMs undergo abrupt reversible gated sorption transitions from an empty collapsed structure to a filled expanded porous state through cooperative adsorption of guest molecules between layer planes. The ELM gating transition and associated hysteresis loop is different than the gas condensation phase transition commonly observed for other nanoporous adsorbents, in that it is induced by rearrangement of the MOF structure rather than by sorbent pore size distribution effects. Certain ELMs have very high selectivity for CO2 adsorption from gas mixtures, and are intriguing adsorbents for cost reductions in carbon capture and storage operations for CO2 removal from nitrogen and other flue gases in post-combustion separations or from hydrogen in pre-combustion separation (e.g. following coal gasification). Moreover, some ELMs have attractive sorption capacities, enthalpies, and isotherm features for efficient recovery of captured CO2 using temperature or pressure swing adsorption. ELMs thus merit further investigation as materials for sorbent-based capture of industrial CO2 emissions.

We report results from molecular simulations of ELM-11 [Cu(BF4)2(bpy)2 (bpy = bipyridine)] and ELM-12 [Cu(bpy)2(OTf)2 (OTf = trifluoromethanesulfonate)] for separation of CO2 from pre- and postcombustion gas mixtures. Adsorption capacity, CO2 selectivity, and isosteric heat of adsorption for pure CO2 and binary CO2/H2 and CO2/N2 mixtures were calculated via grand canonical Monte Carlo (GCMC) simulation at temperatures (273 to 400 K), pressures, and compositions (5 to 15% CO2) representative of power plant flue gas streams. GCMC simulations of CO2 capacity above the gate pressure using a rigid model of ELM-11 agreedwell with experimental isotherms at 273 K. CO2 sorption capacity
(140 mg/g) and heat of adsorption on ELM-11 are competitive with commodity adsorbents at realistic process conditions. Simulations project a CO2/N2 selectivity of 700 at 300 K, a capture ratio significantly higher than common activated carbons and zeolites. The simulated CO2 selectivity is comparable to experimental results for separating CO2 from N2 and O2 by temperature swing adsorption from 268 K to 311 K using ELM-11. Desorption of captured CO2 of more than 99% purity was obtained from a ternary mixture of 40% CO2, 48% N2 and 12% O2 at 268 K. GCMC simulations on ELM-11 for CO2/H2 mixtures representing shifted synthesis gas were also carried out and CO2 selectivities of 1000 and 400 were obtained at 298 and 373 K respectively. These compare favorably with reported values for commercial adsorbents used for gas separations. To investigate framework flexibility effects on CO2 sorption in ELM-11, a hybrid Monte Carlo/molecular dynamics (MCMD) method was applied. CO2 adsorption isotherms showed good agreement with experiments at 273 K.

Mixed-gas simulations yield CO2/N2 selectivities of 70 and 20 respectively at 300 K and 350 K, comparable or superior to other common adsorbents, but substantially lower than selectivities predicted from rigid model framework simulations. Isotherms for CO2 and N2 on ELM-12 are contrasted to those obtained for ELM-11. The bulkier ELM-12 counter-ion yields a structure with microporosity in addition to latent porosity as found in ELM-11. Unlike ELM-11, ELM-12 exhibits a two-step isotherm with micropore filling followed by interlayer clathrate formation at a narrowly defined gating pressure.

See more of this Session: Greenhouse Gases Capture and Sequestration
See more of this Group/Topical: Topical 6: 11th Topical Conference on Gas Utilization

(25f) CO_2 Solubility In Aqueous Amines at High Temperature and Pressure

Qing Xu and Gary T. Rochelle, The Department of Chemical Engineering, The University of Texas at Austin, Austin, TX

CO_2 Solubility in Aqueous Amines at High Temperature and Pressure

Qing Xu, Gary T. Rochelle, gtr@che.utexas.edu

Department of Chemical Engineering

The University of Texas at Austin

Amine based post-combustion CO_2 capture is one of the most effective ways to reduce the CO_2 emission. Various aqueous amines have been investigated by researchers. However previous vapor-liquid equilibrium measurements mostly focused on absorber conditions at 40 to 70 ∞C with an exception of aqueous MEA.

This paper presents total pressure and CO_2 partial pressure data for CO_2 loaded aqueous amines at 100-160 ∞C, 100- 2800 kPa. 7 m (molality, mole/kg water) monoethanolamine (MEA) was chosen as the benchmark and the base line solvent. 8-10 amine solvents with various CO_2 loadings were screened, including MEA, piperazine (PZ), 1-methyl-piperazine (1MPZ), 2-methyl-piperazine (2MPZ), diglycolamine (DGA), dimethyl-piperazine (DMPZ) and blends of selected above amines. These VLE measurements at high temperature contribute to thermodynamics at stripper and reclaimer conditions. Combined with the low temperature data these measurements also give a good interpolated prediction for the mid-temperature range.

The calculated results for 7 m MEA match well with literature. The regressions based on data from 40 to 160 ∞C gave models for CO_2 partial pressure over loaded aqueous amines and the models predict the data well. The CO_2 partial pressure mainly depends on the type of amine, temperature and CO_2 loading (mol CO_2/mol alkalinity). The concentration of amines does not have an obvious effect on CO_2 partial pressure. Heat of absorption for CO_2 loaded aqueous amines was also derived from these models and compared in this paper. Heat of absorption is an important factor in predicting the energy consumption in the stripping process.

(25g) CO_2 Absorption In Blends of Piperazine and Piperazine Derivatives

Xi Chen and Gary T. Rochelle, Chemical Engineering, The University of Texas at Austin, Austin, TX

Piperazine (PZ) is a good amine solvent for CO_2 capture because it has high absorption rate, great cyclic CO_2 capacity and excellent thermal stability. However the application of PZ is limited by the window of solid solubility in water. To accurately screen amine solvents which have better solubility in water, CO_2 solubility and absorption/desorption rate in the blend of PZ and 2-methylpiperazine (2-MPZ) as well as the blend of PZ, 1-methylpiperazine (1-MPZ) and 1,4-dimethylpiperazine (1,4-DMPZ) was measured in a Wetted Wall Column (WWC). The results were compared to those for 8 molal (m) PZ. 4 m PZ / 4 m 2-MPZ was found to have about 5 % higher CO_2 capacity but 15 % lower absorption rate than 8 m PZ. 3.75 m PZ / 3.75 m 1-MPZ / 0.5 m 1,4-DMPZ has a CO_2 capacity 10 % higher than 8 m PZ while retaining the same absorption rate. Quantitative ^{1}H and ^{13}C NMR studies were performed on 8 m 2-MPZ at different CO_2 loading and 40 °C to understand the speciation. It was found that the methyl group greatly hindered the formation of carbamate on the adjacent amino group. The unhindered monocarbamate accounts for more than 80 % of the total carbamate at rich CO_2 loading. With all the experimental data, thermodynamic and kinetic models are under development using Aspen Plus® to represent CO_2 solubility and rate in PZ/2-MPZ and PZ/1-MPZ/1,4-DMPZ.

See more of this Session: Greenhouse Gases Capture and Sequestration
See more of this Group/Topical: Topical 6: 11th Topical Conference on Gas Utilization

(30a) Successful Design and Implementation of An Enhanced Selective Non Catalytic Reduction (ESNCR) System for a Refinery Boiler

Antonio Marquez[1], Wei Zhou[2], David Moyeda[2] and Santosh Nareddy[2], (1)GE Energy, Santa Ana, CA, (2)GE Energy, GE Energy, Santa Ana, CA

Selective non-catalytic reduction technology (SNCR) is an effective and economical method of reducing NO_X emissions from a wide range of industrial combustion systems. It is widely known that the SNCR process is primarily effective in a narrow temperature window, around 900 to 1,000ºC, and that high CO concentrations can both shift the temperature window and limit its effectiveness. However, in a lot of refinery boilers, the flue gas temperature is lower than the normal SNCR temperature. Therefore, in order to design a system that can achieve optimal NO_X reduction performance, careful use of design tools are needed to identify the best location for reagent injection, for maximizing residence time and for minimizing the CO impact. In some cases, additives, such as hydrogen can be used to enhance SNCR (ESNCR) NO_X reduction performance when flue gas temperatures are less than optimal. GE Energy has recently successfully designed and implemented an ESNCR system for a refinery CO boiler and achieved significant NOx reduction (>40%) with less than 10 ppm slip. This paper presents the process design methodology consisting of computational fluid dynamics simulation and chemical kinetics predictions, and discusses the optimization test results. The program demonstrates that the hydrogen enhanced SNCR technology can be effective for refinery boiler NOx control.

See more of this Session: Air Compliance Issues in the Chemical and Refining Industries
See more of this Group/Topical: Environmental Division

(30b) Risk-Analysis Based Flare Load Reduction During Ethylene Plant Upsets

Xiongtao Yang, **Qiang Xu** and Kuyen Li, Dan F. Smith Department of Chemical Engineering, Lamar University, Beaumont, TX

Ethylene plant upsets usually lead to the flaring of off-spec products, resulting in significant raw material and energy losses, as well as air emission problems. Under the premise of plant safe operation, establishing process recycles connecting off-spec streams to their upper-stream process can help ethylene plants reduce flare load during plant upsets. In this paper, recycling two most significant flaring sources, effluents from the acetylene reactor and ethylene tower overhead, are addressed. Operational strategies for recycling the potential flaring sources under various process upsets are developed and analyzed based on rigorous plant-wide dynamic simulations. Risk analysis on the performance of cracked gas compressor (CGC) system has been conducted through the quantitative comparison of the effectiveness of various recycling strategies. Therefore, the study not only discloses how to effectively and efficiently design and operate recycling streams, but also addresses the CGC operating safety. Case studies have demonstrated that the risk-analysis based flare load reduction strategies can proactively reduce the flaring emission amount and upset time, and therefore have great potentials to bring economical and environmental benefits to ethylene plants. The study is helpful in supporting ethylene plant decision-makings on handling plant upsets.

See more of this Session: Air Compliance Issues in the Chemical and Refining Industries
See more of this Group/Topical: Environmental Division

Removal of Aircraft Cabin Ozone by Microfibrous Material at Micro Second Contact Time

Qiang Gu, Chemical Engineering, Auburn University, Auburn, AL and
Bruce Tatarchuk, Chemical Engineering, Auburn University, Auburn, AL

Abstract

Microfibrous Entrapped Catalyst (MFEC) manufactured from 8 μm diameter nickel fibers were engineered into pleated heterogeneous catalytic reactors to improve removal performance from traditional packed bed and monolith reactors, since microfibrous materials can entrap small particles (150-200 μm) which significantly improves inter phase and intra particle mass transfer. These special reactors were targeted at ozone, which was widely recognized as No.1 aircraft cabin air pollutant. In this project, MFEC reactors were investigated under aircraft bleed air conditions of high temperature (150-200C) and high face velocity (10-30 m/s) for different catalysts (Pd, Ag, Mn). Catalysts were impregnated on entrapped particles (e.g. γ-Al2O3) using incipient wetness method. Ozone test concentration was set at a high-demanding 1.5 ppmv. Results showed that even if much less amount of metal catalysts were entrapped in microfibrous materials, a high level of ozone decomposition was accomplished. Compared with conventional aircraft filters, this can be a huge advantage in terms of material cost and labor. Reaction kinetics analyses were compared for different catalysts. Also catalyst aging tests were carried according to the frequency of commercial aircraft ozone reactor replacement, which evaluated long-term performance for actual MFEC reactor usage. In addition, pressure drop improvement by pleating MFEC was also evaluated with comparison of traditional low pressure drop monolith reactors. CFD model has also been established for pressure prediction and comparison. It showed that pleated structure enable MFEC to avoid higher pressure drop and eliminate the problem of direct channeling caused monolith.

1. Introduction

1.1 Microfibrous Material

Microfibrous Entrapped Catalysts (Figure 1) are a special kind of microstructure heterogeneous system. It utilizes the sinter-lock structure of the micron meter fibers to entrap catalyst particles as small as 20 microns. The high voidage structure has improved its performance in terms of pressure drop.

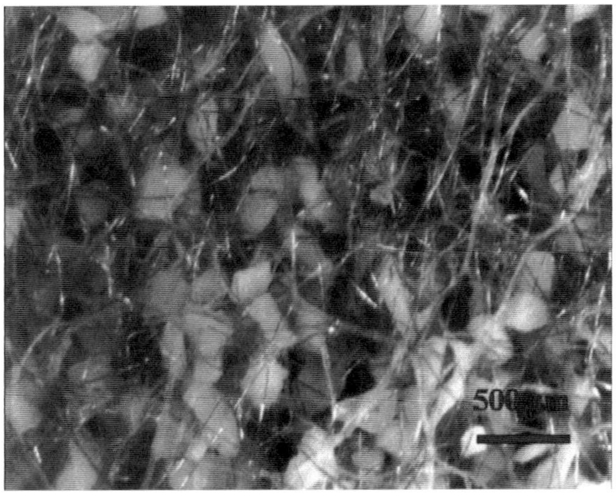

Figure 1 Mircrofibrous entrapped catalysts

In this study, microfibrous entrapped catalysts were engineered into ozone filters, which was majorly aimed at purifying aircraft cabin air. The performance of microfibrous material has been compared with normal packed bed and monolith by Kalluri et al. (2009). Microfibrous material reactors have shown great potential in high volumetric reactivity structures. It eliminates the high pressure drop created by packed bed reactors. Also, channeling in monolith has been avoided, which has led to a better performance in conversion.

1.2 Background of Ozone

Photochemical conversion of oxygen creates ozone in the atmosphere. Ozone is usually found in the earth's upper atmosphere (10~30 miles above the earth's surface). The concentration increases with latitude and changes with seasons and weather. The Global Atmosphere Sampling Program conducted ozone concentration measurement in and out of the cabin of a B-747-100 in 1977 (Figure 2). It showed in Figure 2 that the outside ozone concentration can be well over the limit set by FAA in 1985. According to FAA, cabin air ozone level should not be over 0.25 ppm at any time over 32,000 ft flight. 0.10 ppm should not be passed for any 3-hour flight over 27,000 ft. If the aircraft cabin air has been left untreated, it will cause a series of health problems, including coughing, throat irritation and uncomfortable chest feelings.

Ozone is also created by many common equipments, including copiers, pool cleaners, and water purification processes. Catalytic ozone decomposition has been widely applied to these situations to protect human health. Today, most ozone releasing equipments have installed ozone filters. But some of have been adapted in non-optimized way. For instance, packed bed style converters have been popular in most situations. But in situations like aircraft filter, pressure drop have been a great issue as to packed bed converters. Microfibrous materials filters have been developed as a high contact-efficiency and low pressure drop alternative to these situations.

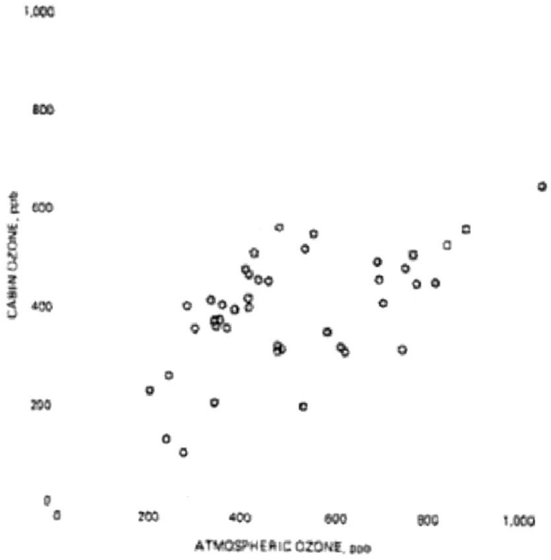

Figure 2 GASP measurements on ozone in and out aircraft cabin

2. Experiment

2.1 Preparation

High surface area alumina was obtained from Alfa Aesar. This alumina was grinded into 150~200 um powders so that they can be entrapped in nickel fibers. Wet-lay process was applied to make microfibrous material sheets. In this process, 8 um diameter nickel fibers was blend into a suspension and was transferred into a head box where alumina powders were mixed with the fibers. The suspension was then drained, leading to a preform of microfibrous material. Preform sheets were dried at 85C overnight before going into the sintering furnace. BTU1000 sintering furnace sinters the sheet at 1000C. The sheets were then impregnate with $Pd(NO_3)_2$ solution. Wet sheets were dried overnight. These sheets were finally calcinated at 450C for 4h. Testing pleat structures were made by aluminum U channels. Sheets were secured within the channels by RTV silicon, which is also used to seal edges in the reactor (Figure 3). Perforated metals were applied at both side of the material sheet to help with strength of the microfibrous material.

Figure 3 Pleated microfibrous material filters

2.2 Setup

Figure 4 shows the scheme of the ozone test loop. A 40hp blower from Fan Equipment Inc. is used to circulate air within the ozone test loop. PLC control is used to change the system speed. System speed is acquired by testing pressure drop across a bare section of the test loop and transferring that to speed by Darcy's Law. These speeds are verified by a commercial pitot tube (Figure 5). Testing velocities range from 10~35 m/s. System temperature is controlled by both purging some portion of the system air and six 1800W heaters on the blower. At high speed, energy crated by skin friction of the test loop is enough to heat up the system to required temperature. Purging air help maintain the temperature. At low speed, heater came into play to heat up the system. Ozone was generated by PZ2-16 from Prozone Water Products and was introduced into the system at the outlet of the air blower. Ozone mixes with air in the test loop to meet the required 1.5ppm. K-type thermo couples were used at upstream and downstream of the reactor to detect system temperature. The tested temperatures are 373K, 423K, 473K.

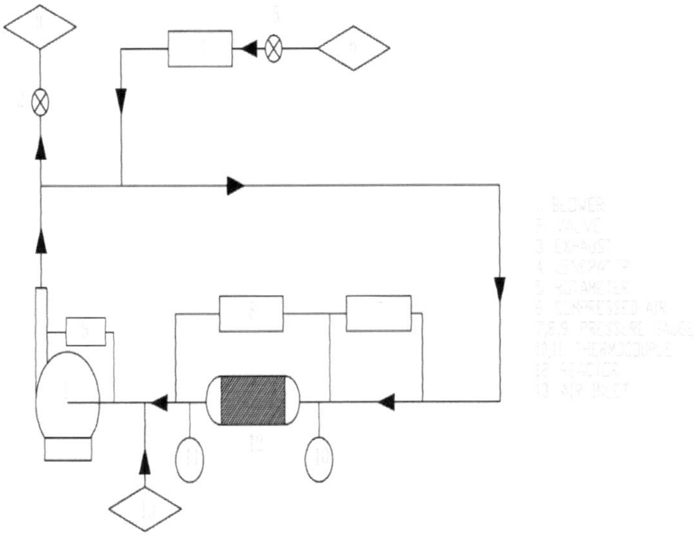

Figure 4 Scheme of ozone test loop

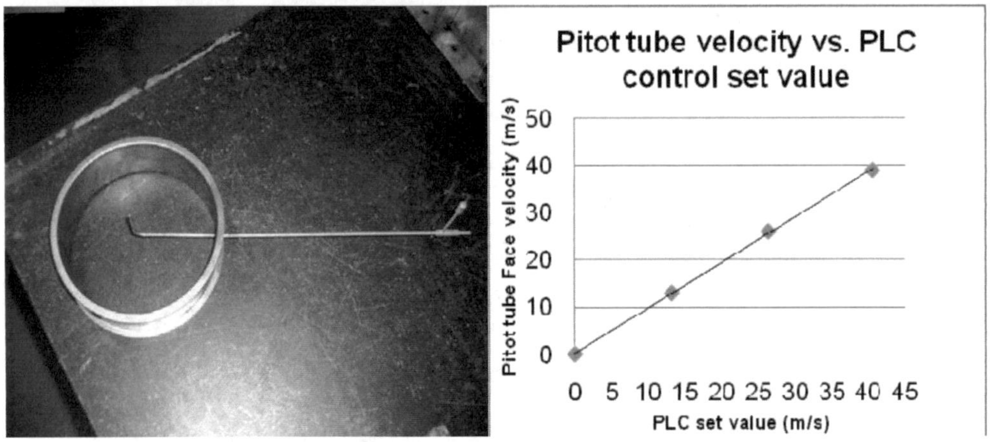

Figure 5 Speed verification of ozone test loop

3. Result and discussion

3.1 Ozone Conversion

Five Ozone Catalysts were tested at 373K, 423K, 473K at face velocities ranges from 10~30 m/s for conversion. Catalysts are listed in table 1.

Catalyst	Weight Percent
Pd/Al_2O_3	1% Pd
Pd/SiO_2	1% Pd
Ag/Al_2O_3	1% Ag
MnO_2/Al_2O_3	10% Mn
$Pd\text{-}MnO_2/Al_2O_3$	1% Pd 10% MnO_2

Table 1 Ozone Decomposition Catalysts

The decomposition of ozone depends on catalyst, contact time and running period. The relationship between face velocities and conversion rates are shown in Figure 6. Linear trends are observed in the figures. Quederni et al. (1996) found that the ozone decomposition over porous media follows the first order reaction (2). We applied the first order plug flow (1) model and Arrhenius (3) Equation to analyze the ozone conversion data.

$$C=C_0e^{-kt} \quad (1)$$

$$r=-\frac{d[O_3]}{dt} \quad (2)$$

$$\ln(k)=\frac{-E_a}{R}\frac{1}{T}+\ln(A) \quad (3)$$

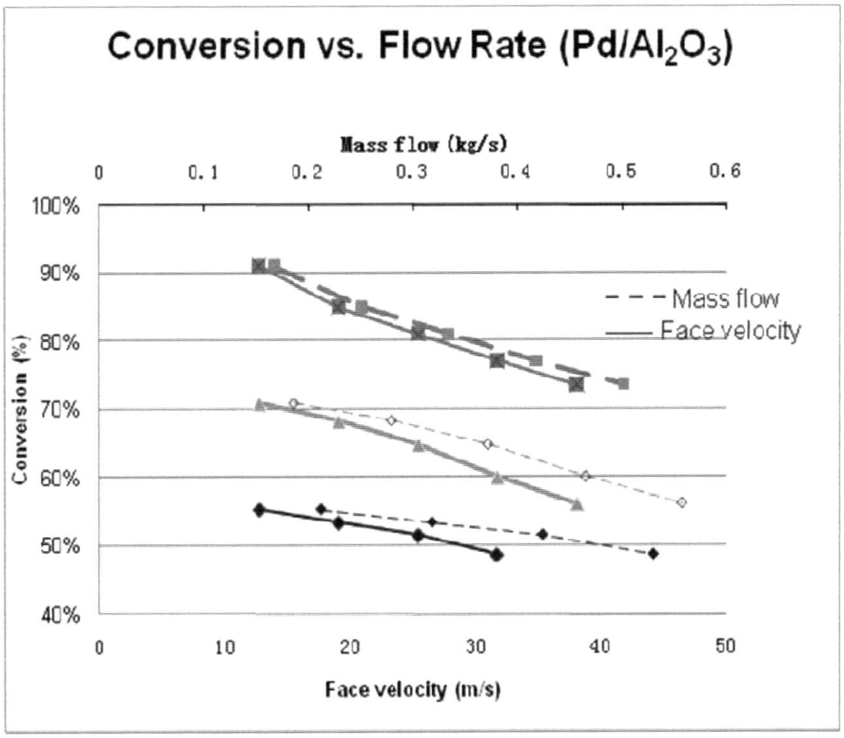

Figure 6 Ozone conversions at different temperatures and face velocities

The Arrhenius fit for 1% Pd/Al2O3 catalyst is shown in Figure 7 as a sample of analysis. Comparison of rate constant and activation energy were listed in table 2. As we observe, activation energy for each catalyst were small, the activity of each catalyst would depend on the catalyst property. Reaction constant k was calculated for both per weight of precious metal and per precious metal atom. As seen, mixture of Pd and MnO_2 has the highest per atom rate constant. This is consistent with its best performance in term of ozone decomposition among catalysts.

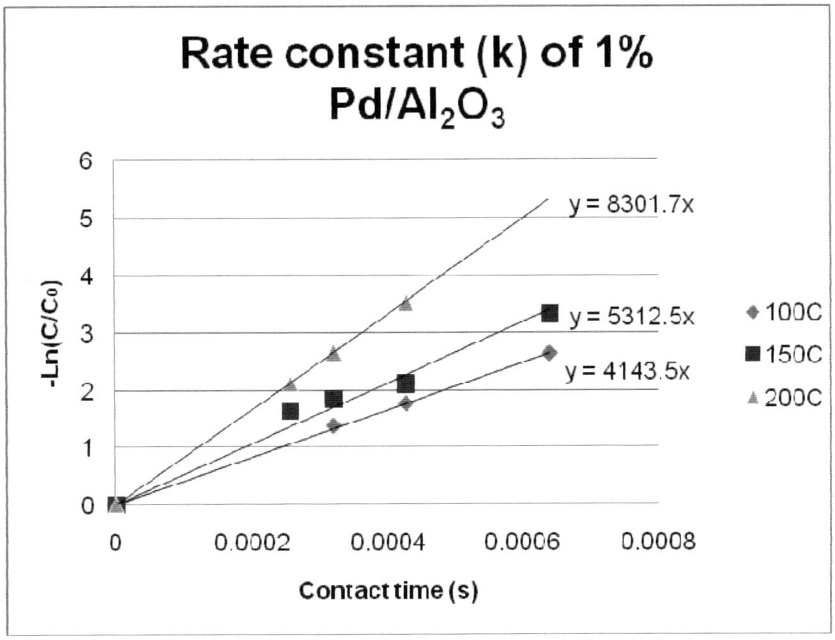

Figure 7 Arrhenius fit for 1% Pd/Al_2O_3

Catalyst	Rate constant k (1/s)			E_a (kJ/mol)	k per wt. of PM (1/s*gram PM)	k per PM atom at 200C (1/s·atom)
	100C	150C	200C			
1% Pd/Al_2O_3	885	1640	4530	23.6	31,701	$5.6*10^{-18}$
1% Pd/SiO_2	574	1463	4403	29.7	30,812	$5.4*10^{-18}$
1% Ag/Al_2O_3	1921	3001	4527	12.5	31,679	$5.7*10^{-18}$
10% MnO_2/Al_2O_3	3801	4304	4710	3.2	3,296	$3.0*10^{-19}$
1% Pd-10% MnO_2/Al_2O_3	4144	5313	8302	10.0	5,810	$7.2*10^{-19}$

Table 2 rate constants and activation energy comparison for ozone decomposition catalysts

3.2 Pressure Drop

Pressure drop has played a significant role in aircraft filter performance. Since lower pressure drop help with energy saving and reduces the operating cost of commercial aircraft, microfibrous material has shown a great improvement compared with packed bed reactors. Kalluri et al. (2009) has performed a comparison between microfibrous material reactors and monoliths. Results showed that microfibrous material has only slight more pressure than monolith. Considering the much better performance in conversion, microfibrous material has shown great capability in aircraft filters. Figure 8 shows the pressure drop at different temperature and speeds on the full size mircofibrous aircraft filters.

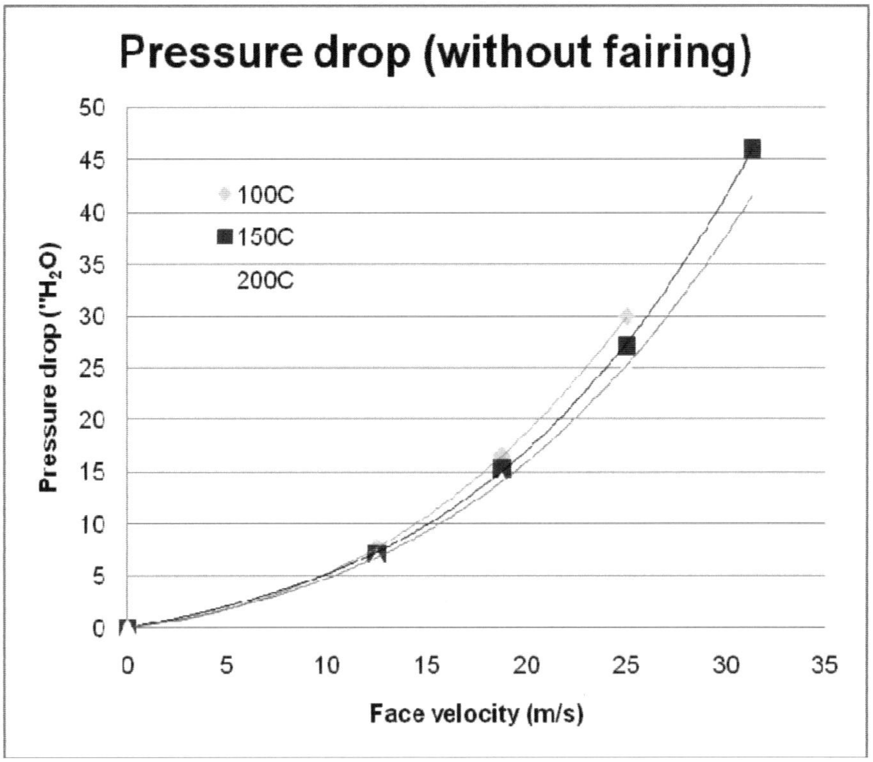

Figure 8 Pressure drop across microfibrous aircraft filters

Summary

In this study, microfibrous material entrapped catalyst filters were tested for aircraft air purification. Catalyst performance was evaluated for five different catalysts. A larger surface area /volume ratio helped with the demand of volume of air handling system in a limited space. Result shows that Pd-MnO_2/Al_2O_3 activity was much higher than other conventional catalysts. Utilizing microfibrous material as aircraft filters also helps with fuel economy. By further engineering design, microfibrous entrapped catalyst system can fit into many compact catalytic systems.

References

Quederni, A., Limvorapituk, Q., Bes, R., and Mora, J. C. *Ozone decomposition on glass and silica. Ozone. Sci. Eng.*, 18, 385–415.

Dhandapani, B., and Oyama, S. T. *Gas phase ozone decomposition catalyst.* Appl. Catal., B, 11, 129–166.

B.J. Tatarchuk, M.R. Rose, A. Krishnagopalan, J.N. Zabasajja, D. Kohler, US Patent 5,080,963 (1992).

B.J. Tatarchuk, M.F. Rose, A. Krishnagopalan, US Patent 5,102,745 (1992).

B.J. Tatarchuk, M.F. Rose, G.A. Krishnagopalan, J.N. Zabasajja, D.A. Kohler, US Patent 5,304,330 (1994).

R. R. Kalluri, Donald R. Cahela, B. J. Tatarchuk, Comparative heterogeneous contacting efficiency in fixed bed reactor: Opportunities for new microstructured system. Appl. Catal., B 90(2009) 507-515

Biodegradation of Phenol-Polluted Air in Aerobic Granular Reactors

Ping Zeng*, Jijun Du**, Yonghui Song***, Liguo Zhao****, Chenhao, Wang*****

*Chinese Research Academy of Environmental Sciences, Beijing, 100012, PR China. Phone: 86-10-84919726-602 E-mail: zengping@ntu.edu.sg

**Chinese Research Academy of Environmental Sciences, Beijing, 100012, PR China. Phone: 86-10-84934516-8004 E-mail: xiaodu2000@sina.com

***Chinese Research Academy of Environmental Sciences, Beijing, 100012, PR China. Phone: 86-10-84928380 E-mail: Songyh@craes.org.cn

****School of Civil and Environmental Engineering,University of Science and Technology Beijing, 100083, PR China. 86-10-84911236 E-mail: zhaoliguo1011@gmail.com

****China Shipbuilding Ninth Design & Research Institute Engineering Co.,Ltd, Shanghai, 200063 , PR China. 86-21-62549700-779 E-mail: wangchenhao2006@gmail.com

ABSTRACT

As one kind of toxic volatile organic compound (VOC), phenol-polluted air was degraded by aerobic granules. Aerobic granulation is a novel biotechnology, which has a strong ability to tolerate toxicity and to handle high phenol loading rates. The objective of this work is to evaluate the possibility of phenol elimination by aerobic granules and the characteristics of the aerobic granules. In this experiment, phenol was added to the reactor at loading rates of 0.5, 1.0, 1.5 and 2.0 kg/m^3·d. Aerobic granules showed good degrading ability at different phenol loading rates. At a steady state, the phenol removal rate was higher than 96% while the COD removal rate was higher than 95%. The phenol loading rate exerted a profound influence on the activity, morphology, and structure of the aerobic granules. The specific oxygen utility rate of the aerobic granules changed from 91 to 32 mg O_2/gVSS·h when the reactor was fed with phenol-polluted air. Then it increased to 74 mg O_2/gVSS·h at phenol loading rate of 1.5 kg/m^3·d. Extracellular polymers, as well as protein, polysaccharides increased with the increasing of phenol loading rates. The sludge volume Index (SVI) and settling velocity were 42.5-44 ml/g and 72.6-80.0 m/h, respectively. The compact structure of the aerobic granules was maintained after fed with phenol waste gas. This study confirmed the application of aerobic granule technology on VOCs elimination.

KEYWORDS:

aerobic granules; phenol-polluted air; biodegradation; microbial activity; ECPs;

INTRODUCTION

Volatile organic compounds are a group of organic chemicals which have a boiling point below 260°C at ambient pressure or saturated vapor pressure above 71 Pa at ambient temperature (de Nevers, 1995). Phenol is a volatile organic compound (VOC) emitted by the chemical process industries, which manufacture many phenolic resins and organic solvents (Wallace et al., 1996). The major sources of phenol containing wastewater and waste air streams are petroleum refineries, petrochemical, steel mills, coke oven plants, coal gas, synthetic resins, pharmaceuticals, paints, plywood industries and mine discharge (Mukherjee et al., 2007). Phenol was listed in the priority list of hazardous substances (ATSDR, 2005) since it was regarded as a Hazardous Air Pollutant under the US Environmental Protection Agency Clean Air Act section 112(b) (USEPA, 2006). With the increasingly stringent environmental regulations and the focus on the significant health and environmental problems caused by the release of VOCs (Devinny et al., 1999; Shareefdeen and Singh, 2005), including phenol, major efforts are required to prevent the release of these compounds into the atmosphere and/or to eliminate them from waste air streams (Moussavi and Mohseni, 2008).

Several techniques for VOCs elimination that have been investigated such as thermal incineration, catalytic oxidation, condensation, absorption, adsorption, and membrane separation and other nonbiological treatment methods suffer from serious drawbacks such as high cost and formation of hazardous by-products (Loh et al. 2000; Citepa, 2005). Biological degradation is generally preferred due to easy design and maintenance (low capital and running costs) and being environmental friendly with producing harmless secondary byproducts (Daubert et al., 2001). However, the conventional treatments of phenol gas streams have inherent limitation such as low tolerance to the high phenol loading and failure to adapt to the fluctuations in phenol loads (Watanabe et al. 1996, 1999; Kibret et al. 2000).The innovation of more feasible and more cost-effective technology with a strong resistance to the phenol toxicity, , and to be usedfor phenol elimination is required.

Aerobic granulation, a new form of cell immobilization for exploitation in biological wastewater treatment, has shown a strong performance in toxic wastewater such as phenol wastewater and high concentration wastewater treatment (Beun et al., 2000; Tay et al., 2001; Jiang et al., 2002; Moy et al., 2002; Dulekgurgen, 2003; Arrojo et al., 2004; McSwain et al., 2005). Thus it had the potential to degrade phenol in the air stream. In the present study, a sequencing batch reactor was set up to study the degradation of phenol-polluted air by aerobic granules and the influence of the phenol loading rate on the characteristics of aerobic granules.

METHODOLOGY

Inoculum

The matured acetate aerobic granules which were obtained from a sequencing batch reactor (SBR) were used as seed sludge. They were cultivated from acclimated activated sludge. Acetate was fed as the sole carbon and energy source with a synthetic wastewater with the following composition (L^{-1}): 0.8 g NH_4Cl, 0.045 g K_2HPO_4, 0.03 g $CaCl_2{\cdot}H_2O$, 0.025 g $MgSO_4{\cdot}7H_2O$, 0.02 g $FeSO_4{\cdot}7H_2O$ and micronutrients, which had been described previously (Zeng et al., 2007). Organic loading rate was 3 kg $COD/m^3{\cdot}d$. The characteristics of the inocula are listed in Table 1.

Table 1 Characteristic of aerobic granular sludge

Parameter	Value
MLSS (g/L)	14.4
MLVSS/MLSS (%)	69.1
Mean biomass size (mm)	3.1
Settling velocity (m/h)	76.7
SVI (ml/g)	43.5
COD removal efficiency (%)	95.4
SOUR ($mgO_2/gVSS{\cdot}h$)	91.7
EPS (mg/gVSS)	5.51

Experiment Design and Set-up

Experiments were performed in a column-type sequence batch reactor (SBR) with a working volume of 2.4 L and a diameter of 5 cm. Reactors were operated sequentially in 4-h cycles (4 min of influent filling, 228 min of aeration, 5 min of settling and 4 min of effluent withdrawal) with a hydraulic residence time (HRT) of 8 h. Effluent was discharged at a volumetric exchange ratio of 50%. Fine air bubbles for aeration were supplied through a dispenser at the reactor bottom at an airflow rate of 3.5 L/min and air flow was controlled by a gas flow meter. Phenol waste gas was fed as the sole carbon source with the phenol loading rate increasing gradually from 0.5 to 2.0 kg/m^3·d.

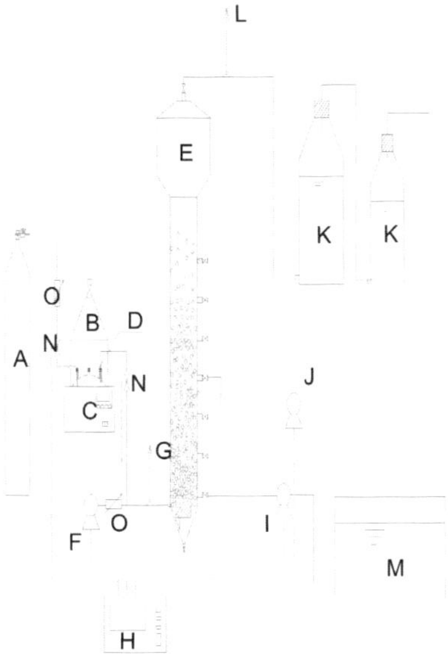

A: high pressure nitrogen cylinder, B: insulating layer, C thermostat water bath, D: flask、E: SBR reactor, F: air pump, G: inlet gas sampling pot, H: timer、I: influent pump, J: effluent pump, K: off-gas washing bottle, L: exhaust sampling pot, M: influent tank, N: control valve, O: antisepticise gas flow-meter.

Figure 1 The image of aerobic granular system.

Medium

The synthetic phenol waste gas combined with nitrogen gas was pumped into the reactor by a dispenser in the bottom of the reactor, to provide the sole carbon source. The concentration of phenol gas was 10.85 mg/l, which provided an phenol loading rate of 0.5 kg/m^3·d. The composition of the micronutrient solution fed into the SBR was same as the one cultivated inocula aerobic granules.

Analytic methods

Suspended solids (SS), volatile suspended solids (VSS), sludge volume index (SVI) and dissolved oxygen (DO) were measured periodically according to the standard methods (APAH, 1998). The specific oxygen utilization rates (SOUR), settling velocity were measured as previously done (Tay et al., 2001, 2005). Extracellular polymers (ECPs) in the granules were extracted by centrifuge and heating extraction method (Jiang et al., 2002). The carbohydrate content in the ECPs was measured using the phenol-sulfuric acid method (Dubois et al., 1965), and the protein content was measured using the bicinchoninic acid protein assay (stoscheck, 1990). Phenol gas was absorbed by 0.1 mol/L NaOH using porous plate equipment and measured as previously done (Jiang et al., 2002). Each determination was performed in triplicate.

Specific phenol utilization rates (SPURs) were obtained by fitting the timed disappearance of phenol in the reactor to a zero-order model that is typically used for biodegradable substrates (Eckenfelder, 2000). The rate was determined by the equation:

$$SPUR = (S_i - S_f)/TX$$

where S_i and S_f are the initial (or start-of-cycle) and final concentrations of phenol (mg l^{-1}), respectively, T is the time of phenol disappearance (h) and X is the dry weight of the granules (g l^{-1}).

Granule morphology was determined by COIC microscope and OLYMPUS camera (SP-350). The sizes were measured by a laser particle size analysis system (Malvern Mastersizer 2000) and analyzed by image analysis system

(Image-Pro Plus 5.0). Microscopic study of the aerobic granular samples was conducted with a scanning electron microscope (Stereoscan, 420, Leica Cambridge Instruments), as described previously (Zeng et al., 2007).

RESULTS AND DISCUSSION

In this experiment, the reactor was operated through four stages with phenol loadings rate ranging from 0.5 to 2.0 kg phenol/m^3·d with inlet phenol waste gas concentrations ranging from 10.9 to 43.4 mg/L. After about 2 months of operation, all four reactors reached a steady state, as evidenced by stable biomass concentrations and constant phenol removal efficiencies.

Phenol Waste Gas Degradation Profiles at Different Phenol Loading Rates

Aerobic granules were first cultivated in sequencing batch reactors with sodium acetate as sole carbon source. The long-term performance of the aerobic granular reactor was investigated with various inlet phenol concentrations ranging from 10.85 to 43.4 mg/L, corresponding to loading rates of 0.5-2.0 kg phenol/m^3·d. This section presents the mean phenol removal efficiency of the granular reactor based on data gathered under steady state condition. The steady state was assumed when the phenol removal efficiency remained constant (with changes of less than 3%) over a week period.

Phenol-polluted air was then added to the reactor at phenol loading rates of 0.5, 1.0, 1.5 and 2.0 kg/m^3·d. The relationship between the stable biomass concentration in the reactor and the phenol loading is shown in Table 1. The biomass concentrations increased with phenol loading to peak at a phenol loading rate of 1.5 kg/m^3·d, and declined thereafter. Aerobic granules showed good degrading ability to phenol with the COD removal rate higher than 95% and phenol removal rate higher than 96% (Table 2). The highest COD removal rate was 96.5% at a phenol loading rate of 1.5 kg/m^3·d.

The aerobic granular reactor showed a strong performance on the degradation of phenol waste gas. The optimal phenol loading rate was 1.5 kg/m^3·d based on biomass concentration, the COD and phenol removal rate. For the phenol wastewater, the optimal phenol loading rate was 2.0 kg/m^3·d, which was higher than dealing with phenol waste gas (Jiang et al, 2004). The greater performance of the aerobic granular in this study could be due to the presence of a mobile liquid phase that assisted the mass transfer of phenol (with low Henry's Law constant) from the gas phase to liquid. In addition, the active population of phenol degrading bacteria in aerobic granules played an important role in effectively metabolizing the transferred contaminant.

Table 2 Reactor performance at different phenol loading rates

Phenol loading rate(kg/m^3·d)	**0.5**	**1.0**	**1.5**	**2.0**
Inlet phenol concentration(mg/L)	10.9±0.5	21.7±1.1	32.6±1.6	43.4±2.2
MLSS (mg/L)	6.72±0.34	8.26±0.42	9.06±0.45	8.93±0.44
COD removal rate(%)	95.2±0.5	96.3±0.5	96.5±0.5	96.2±0.5
Phenol removal rate (%)	96.35±0.48	98.27±0.49	98.67±0.49	98.7±0.49

Effects of Phenol loading on Activity of Granules

Figure 2 shows the profiles of specific phenol utility rate (SPUR) at different phenol loading rates. The SPUR was an important index of the phenol degradation capability of microorganisms in aerobic granules (Tay et al., 2005). It means the phenol degradation rate by per unit microorganisms. The SPUP increased from 0.49 mg phenol/gVSS·d at phenol loading rate of 0.5 kg/m^3·d to 1.5 mg phenol/gVSS·d at phenol loading rate of 1.5 kg/m^3·d. Finally, it increased to 1.58 mg phenol/gVSS·d at phenol loading rate of 2.0 kg/m^3·d. The SPUR of aerobic granules increased with the increasing phenol loading rate. It indicated that the activities of aerobic granules were not inhibited by phenol when the phenol loading rate did not exceed 2.0 kg/m^3·d.

The aerobic granular reactor showed good degradation capability when degrading phenol waste gas. The SPUR increased with the increasing of phenol waste gas loading rate. It indicated that there was no toxic inhibition on the aerobic granules before the phenol loading rate exceeded the phenol loading rate of 2.0 kg/m^3·d.

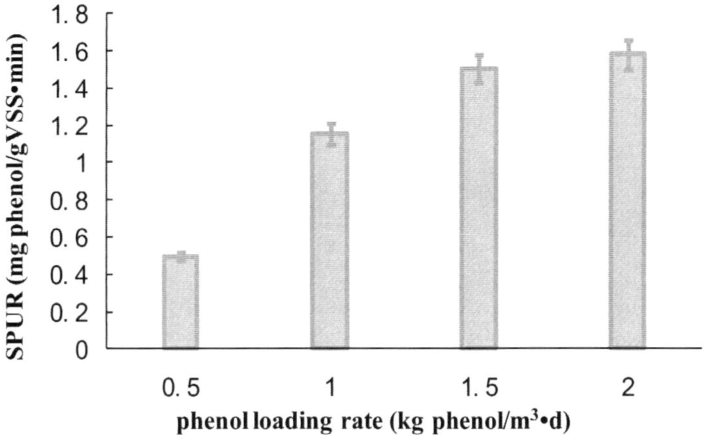

Figure 2 Profiles of specific phenol utilization rates (SPUR) at different phenol loading rates

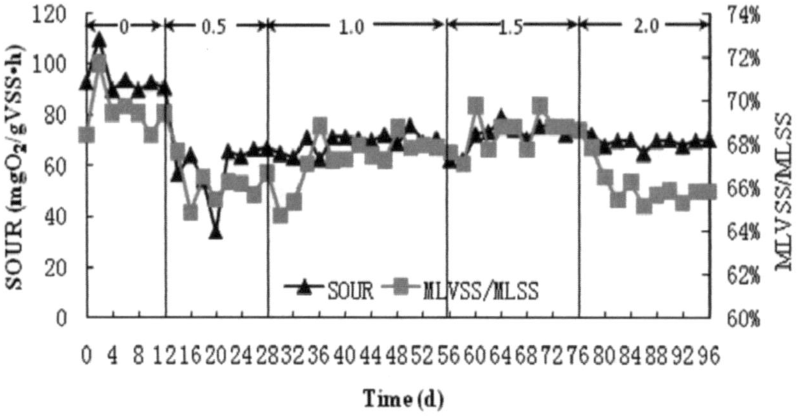

Figure 3 The profiles of SOUR and MLVSS at different phenol loading rates

Figure 3 also shows the effect of phenol loading on SOUR of granules. Biomass growth and substrate removal are linked to batch measurement of SOUR. The SOUR of seed sludge was 90.9 mg O_2/gVSS·h. It decreased to 66.8 mg O_2/gVSS·h at a phenol loading rate of 0.5 kg/m^3·d. It was 62.5 mg O_2/gVSS·h at a phenol loading rate of 1.0 kg/m^3·d. After that, the SOUR of aerobic granules increased to 75.5 mg O_2/gVSS·h at phenol loading rate of 1.5 kg/m^3·d and decreased to 70.3 mg O_2/gVSS·h at phenol loading rate of 2.0 kg/m^3·d. In summary, SOUR peaked at a phenol loading of 1.5 kg/m^3·d, and declined thereafter.

The ratio of MLVSS to MLSS had the similar change trends. It decreased to 66.67% at phenol loading rate of 0.5 kg/m^3·d and increased to 68.64% at phenol loading rate of 1.5 kg/m^3·d. Then it decreased to be 65.80% at phenol loading rate of 2.0 kg/m^3·d.

The metabolic characteristics of the granules followed the same trend as the aerobic granular reactor performance over the range of phenol loadings tested. The highest SOUR was achieved at the phenol loading of 1.5 kg/m^3·d, which was lower than that when degrading phenol wastewater.

Effect of Phenol Loading on Structure of Granules

Figure 4 shows the effect of loading on ECPs of granules, while Figure 5 shows the effect of phenol loading on the ratio of polysaccharides (PS) to proteins (PN) in ECPs.

The amount of PS, PN and ECPs peaked at a phenol loading of 1.0 kg/m^3·d. When the phenol loading rate was lower than 1.0 kg/m^3·d, the amount of PS was a little higher than the amount of PN. When the phenol loading rate was higher than 1.0 kg/m^3·d, the amount of PS was almost 2 times the amount of PN. An increase in phenol loading from 0.5 to 1.0 kg/m^3·d was associated with a 5.5-fold increase in the amount of PN in ECPs, from 9.4 to 47.7 mg/gVSS, while the amount of PS increased 5.0-fold, from 7.0 to 38.4 mg/gVSS. As a result, the amount of ECPs in granules at phenol loading rate of 1.0 kg/m^3·d was much larger than those at phenol loading rates of 0.5, 1.5, 2.0 kg/m^3·d, but the highest PS/PN ratio was 2.6 at the phenol loading rate of 1.5 kg/m^3·d.

Similar to the aerobic granules degrading phenol wastewater (Jiang et al., 2003), ECPs not only act as construction materials for microbial aggregates and are responsible for their structural integrity, but also form a protective shield for the cells against the adverse influences of the external environment. The relationships among PS, PN and the other macromolecules composed of ECPs determine the stability of whole aerobic granules (Sutherland 2001). When the phenol loading rate changed from 0.5 to 1.5 kg/m^3·d, there was no obvious differences between the productions of PN, the increasing of PS/PN was tied to the increasing production of PS. To gain more protection, the microorganisms in aerobic granules were stimulated to produce more PN when they encountered phenol loading rate of 2.0 kg/m^3·d. As a result, the PS/PN decrease corresponded to decreasing specific gravities and increasing SVI values. The stabilities of aerobic granules decreased.

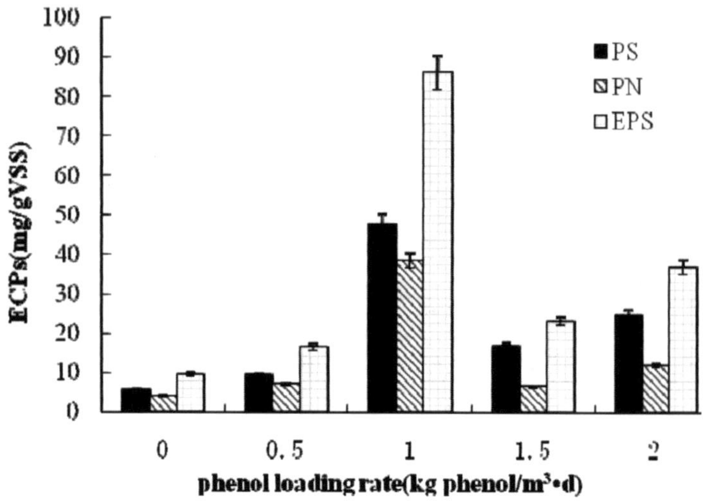

Figure 4 Effect of phenol loading on the amount of polysaccharides (PS), proteins (PN) and extracellular polymers (ECPs)

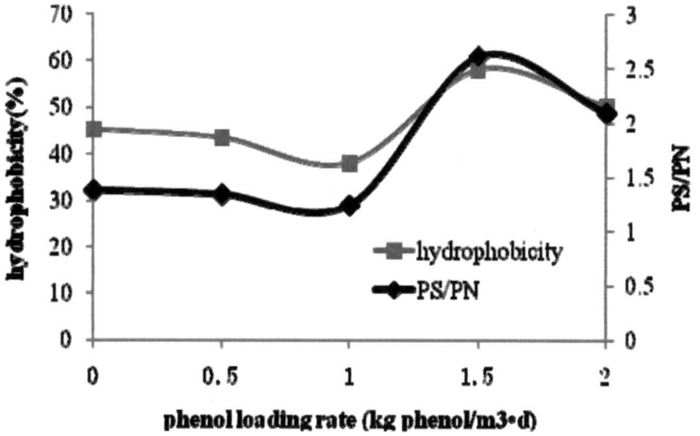

Figure 5 Effect of phenol loading on the ratio of PS to PN

The Cell hydrophobicity plays an important role in the self-immobilization and attachment of cells to a surface. The cell hydrophobicities at different phenol loading rate were tested and shown in Fig. 5. The hydrophobicity of granules was around 38.1-43.5% at phenol loading rate of 0.5, 1.0 kg/m^3·d. It increased to 58.1% at phenol loading rate of 1.5 kg/m^3·d, finally dropped to 50.3% at phenol loading rate of 2.0 kg/m^3·d.

The comparison of the morphology and structure of aerobic granules before and after phenol waste gas fed into reactor was shown in Figure 6. The inocula aerobic granules had a smooth outer surface with a compact and regular appearance. The interior structure was loose. The granules fed with phenol waste gas at phenol loading rate of 1.5 kg/m^3·d had a coarse surface with a compact appearance. The interior of the aerobic granules was dominated by cocci and rod-shaped bacteria that clustered tightly together in an extracellular polymeric matrix.

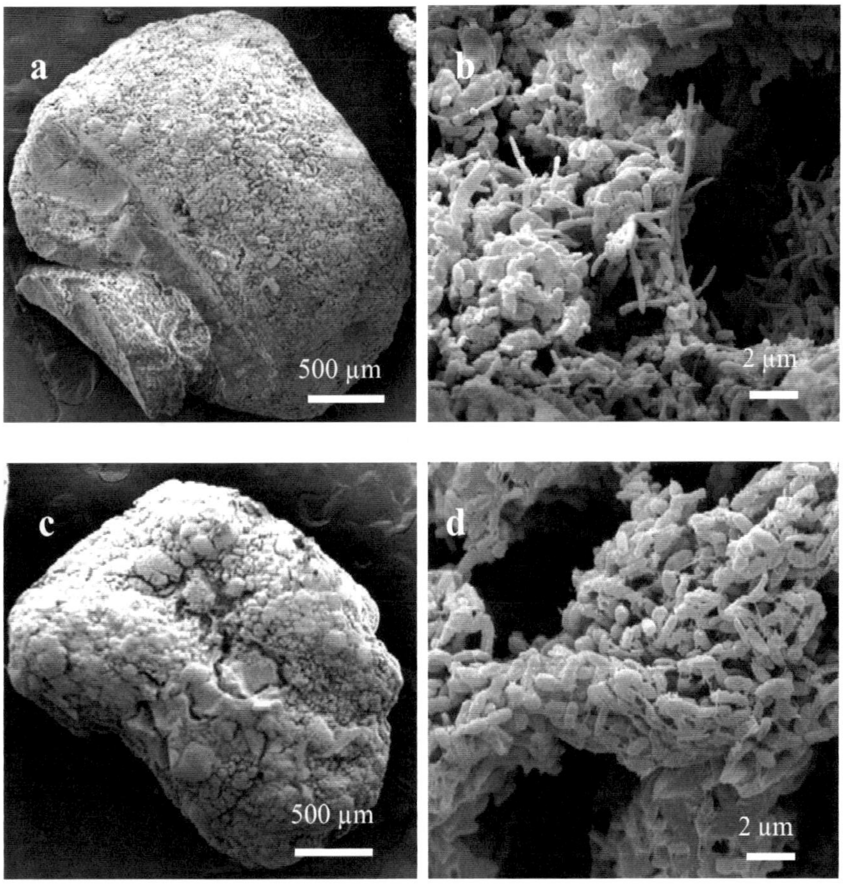

Figure 6 Scanning electron micrographs of aerobic granules, (a) Inocula aerobic granules (b) Surface of inocula aerobic granules (c) Phenol degrading aerobic granules at phenol loading rate of 1.5 kg/m^3·d (d): Surface of phenol degrading aerobic granules at phenol loading rate of 1.5 kg/m^3·d

CONCLUSIONS

A laboratory scale sequencing batch reactor, enriched with phenol degrading aerobic granules, was investigated for the biodegradation of phenol in contaminated gas streams. The aerobic granular reactor indicated effective phenol degradation at different phenol loading rates (0.5–2 $kg/m^3 \cdot d$). The removal of phenol varied between 96.35 and 98.7% depending on the inlet concentration. The SPUP increased from 0.49 to 1.58 mg phenol/gVSS·d when phenol loading rate increased from 0.5 to 2.0 $kg/m^3 \cdot d$. The activities of the aerobic granules were not inhibited by phenol when the phenol loading rate did not exceed 2.0 $kg/m^3 \cdot d$. The aerobic granules showed good structure with the increase of PS/PN from 1.2 to 2.6. Considering the effect of phenol degradation rate, biomass activity and structure of aerobic granules, the aerobic granular reactor performed best at phenol loading rate of 1.5 $kg/m^3 \cdot d$. The results obtained from this investigation provide comprehensive knowledge to better understand the behavior of the aerobic granular reactor treating phenol under actual operational conditions and create the operational procedure to reduce the release of this contaminant under such conditions.

ACKNOLEDGEMENTS

This research was financially supported by National Public Fundamental Funding from Chinese Research Academy of Environmental Science (No.2007KYYW12), the National Key Project of Water Pollution Control (No. 2008ZX07208-003) and the National Natural Science Foundation of China (No. 50905281 and No. 50708101).

REFERENCES

APHA. (1998). The standard methods for the examination of water and wastewater (20th ed.) Washington, DC, USA: American Public Health Association.

Arrojo B., Mosquera-Corral A., Garrido J. M. and Mendez R. (2004). Aerobic granulation with industrial wastewater in sequencing batch reactors, *Water Research.*, **38**(14-15), 3389-3399.

APHA. (1998). The standard methods for the examination of water and wastewater (20th ed.) Washington, DC, USA: American Public Health Association.

ATSDR, (2005). CERCLA priority list of hazardous substances. Available at: http://www.atsdr.cdc.gov/cercla/05list.html.

Beun J. J., van Loosdrecht M. C. M. and Heijnen J. J. (2000). Aerobic granulation, *Water Science and Technology,* **41**(4-5), 41-48.

Citepa website (2005): http://www.citepa.org/pollution/source_en.htm#cov,

Daubert I., Lafforgue C.; Morarges C. and Forade C. (2001). Feasibility study of a compact process for biological treatment of soluble VOCs polluted gaseous effluent. *Biotechnology Progress*, **17(6)**, 1084-1092
Devinny J. S., Deshusses M. A., Webster T. S. (1999). Biofiltration for Air Pollution Control. CRC-Lewis Publishers, New York, USA.
Dubois M., Gilles K. A., Hamilton J. K., Rebers P. A. and Smith F. (1956). Colorimetric method for determination of sugars and related substances, Analytical Chemistry, **28(3)**, 350-356.
Dulekgurgen E., Ovez S., Artan N. and Orhon D. (2003). Enhanced biological phosphate removal by granular sludge in a sequenicng reactor. *Biotechnology Letters,* **25(9),** 687-693.
Eckenfelder, W. W., Jr. (2000), Industrial water pollution control. New York: McGraw-Hill Book Company.
Jiang H. L., Tay J. H. and Tay S. T. L. (2002). Aggregation of immobilized activated sludge cells into aerobically grown microbial granules for the aerobic biodegradation of phenol. *Letters of Applied Microbiology*, **35**, 439-445.
Jiang H. L., Tay J. H. and Tay S. T. L. (2004). Changes in structure, activity and metabolism of aerobic granules as a microbial response to high phenol loading, Applied Microbiology and Biotechnology, **63(5)**, 602-608.
Kibret M., Somitsch W. and Robra K. H. (2000). Characterization of phenol degrading mixed population by enzyme assay. *Water Research,* **34**, 1127-1134.
Loh K. C., Chung T. S. and Ang W. F. (2000). Immobilized-cell membrane bioreactor for high-strength phenol wastewater. *Journal of Environmental. Engineering,* **126(1)**, 75-79.
Moussavi G. and Mohseni M. (2008). The treatment of waste air containing phenol vapors in biotrickling filter. *Chemosphere*, **72**, 1649-1654
Moy B. Y. P., Tay J. H., Toh S. K., Liu Y. and Tay S. T. L. (2002). High organic loading influences the physical characteristics of aerobic sludge granules. *Letters in Applied Microbiology,* **34(6),** 407-412.
Mukherjee S., Kumarb S., Misra A. K. and Fa M. (2007). Removal of phenols from water environment by activated carbon, bagasse ash and wood charcoal. *Chemical Engineer Journal,* **129(1-3)**, 133-142.
McSwain B. S., Irvine R. L., Hausner M. and Wilderer P. A. (2005). Composition and distribution of extracellular polymeric substances in aerobic flocs and granular sludge. *Applied Environmental Microbiology,* **71,** 1051-1057.
Noel de Nevers (1995). Air Pollution Control Engineering. McGraw-Hill, Inc.
Shareefdeen Z. and Singh A. (Eds.) (2005). Biotechnology for Odor and Air Pollution Control. Springer-Verlag, Heidelberg, Germany.
Stoscheck C. M. (1990). Quantitation of protein, Methods in Enzymology, **182**, 50-68.
Sutherland I. W. (2001). Biofilm exopolysaccharides: a strong and sticky framework. *Microbiology* **147**: 3-9
Tay J. H., Liu Q. S. and Liu Y. (2001). Microscopic observation of aerobic granulation in sequential aerobic sludge blanket reactor, *Journal of Applied Microbiology,* **91,** 168-175.
Tay S. T.-L., Moy B. Y.-P., Jiang H. L. and Tay J.-H. (2005). Rapid cultivation of stable aerobic phenol-degrading granules using acetate-fed granules as microbial seed. *Journal of Biotechnology*, 115(4): 387-395.
Tay J.-H., Liu Q.-S. and Liu Y. (2001b) The effects of shear force on the formation, structure and metabolism of aerobic granules. *Applied Microbiology and Biotechnology*, 57(1-2): 227-233.
Wang Z., Liu L., Yao J. and Cai W. (2006). Effects of extracellular polymeric substances on aerobic granulation in sequencing batch reactors, *Chemosphere*, 63(10): 1728-1735
USEPA, (2006). Hazardous air pollutants. Clean Air Act. U.S. Environmental Protection Agency, Washington, DC. United States Code 42 USC 7412. http://www.epa.gov/ttn/atw/orig189.html

Wallace J. C., Basu I. and Hites R. A., (1996). Sampling and analysis artifacts caused by elevated indoor air polychlorinated biphenyl concentrations. *Environmental Science & Technology*, **30(9)**, 2730-2734.

Watanabe K., Hino S. and Takahashi N. (1996). Responses of activated sludge to an increase in phenol loading. *Journal of Fermentation and Bioengineering,* **82(5)**, 522-524.

Watanabe K., Teramoto M. and Harayama S. (1999). An outbreak of nonflocculating catabolic populations caused the breakdown of a phenol-digesting activated sludge process. *Applied and Environmental Microbiology,* **65(7)**, 2813-2819.

Zeng P. , Zhuang W.-Q. , Tay T.-L. Stephen , Tay J.-H. (2007). The influence of storage on the morphology and physiology of phthalic acid-degrading aerobic granules , *Chemosphere* **69(11)**: 1751-1757

REAL TIME CO_2 AND ENERGY COST REDUCTIONS: FIELD EXPERIENCES IN PETROCHEMICAL PLANTS

Carlos Ruiz, Oscar Santollani, Soteica Ideas & Technology LLC, Houston, TX

Introduction

CO_2 emissions are in most part by-products of utility systems; therefore it is only logical to model emissions along with the energy system. Any attempt to reduce the costs from the utility side (i.e.. to optimize the energy system) should be based on the appropriate utility system models (calibrated with validated and consistent set of measurements) and a compositional fuel representation to allow combustion-based Carbon calculations. Solving and optimizing the energy and Carbon balances together is key to be consistent with operational and constraints handling.

The optimization's goal should be to reduce the overall utilities cost taking into account the constraints associated with the existing equipment, fuels/power pricing and contracts. The economic optimization should be done by considering the emission limits as constraints and/or part of the objective function (for example, cost or credit of the CO_2 emission if above or below a certain quota). The energy management system models need to be executed and optimized at a scheduled frequency with online, real time data.

In addition to the optimization, Key Performance Indicators (KPIs) should also be calculated and stored in the Site Plant Information System or DCSs for Operations and Management use.

However, like in any other human activity, the technology by itself is not sufficient to guarantee the success of an energy and CO_2 reduction initiative. Over a period of more than 20 years of deploying industrial projects related to energy systems real time optimization and cost reduction in refining and petrochemical industries around the world, the following or similar issues have been found:

- Many applications of energy cost and CO_2 reduction technologies are successful when used offline by Engineers, at their offices, but only a few of them really succeed when transferred to the Operators to be routinely used in real time, tied to online field data, under unsupervised automatic execution.

- In some cases, systems that are deployed after successful, well engineered, real time projects and that are initially received enthusiastically by Engineers and Operators are only used for a few months after project completion and then fall in disuse.

- Real time systems with acceptable initial up-time figures end up degrading and finally are left offline, unused, after the Engineer in charge is promoted to a new position as a consequence of being so successful with them initially.

- Very well proven, sound technologies, fail when applied at a given Site while they succeed at a different location owned by the same Company.

- After a failure with one of those advanced technologies, the Site or the Company becomes “immunized” against similar technologies. Only after several years and with the renewal of the Engineering staff does the technology have a chance to be considered again.

This paper addresses some of the root causes for the above mentioned issues and discuss the best practices that will help to avoid project failures. Main project steps are explained and critical details to be taken into account to assure successful use and proper technology transfer are discussed. It also presents real industrial examples in which the whole utilities system cost of a production Site (i.e., steam, fuels, boiler feed water and electricity) is optimized with a real time, online, industrially well established software.

Real Time Energy Management Systems

Refineries and petrochemical plants usually operate large and complex utilities and energy systems. They burn several type of fuels, operate cogeneration units to supplement the electric power purchases or export electricity, need to provide several steam pressure levels to different types of consumers and observe emission limits.

In almost all of the cases, these complex utilities and energy systems have several degrees of freedom. Proper manipulation of the degrees of freedom with the aid of cost-based optimization software can result in significant operating cost savings. Since the electrical system represents one of the main economic trade-offs with the steam system and Site Carbon emission, electrical deregulation provides many new challenges to operate the overall combined energy system at minimum cost.

Other important aspects are that utilities systems are continuously evolving (changes are frequent) and that sometimes, there is a lack of sensors and often measurement errors need to be addressed properly.

In addition, utilities systems have several constraints that usually come from the operational and contractual sides. For example, minimum and maximum steam production capacity in the boilers, NO_x / CO_2 emission constraints, the need to maintain a steam production cushion due to safety reasons, or the fuels and electric power quotas and penalties.

Also, it is important to mention that traditionally, given the complexity of the systems, the optimization of the utilities was traditionally managed at the level of each individual Plant or Area. But the optimization of each individual Area does not necessarily give the true overall site optimum. Moreover, an attempt to optimize a whole site based on an isolated view of each Plant at a time could be worse than doing nothing.

In the dynamic operational and economical environment of a Plant, implementing the proper actions necessary for the optimum utilities system management can only be done with the aid of an on-line tool. It should execute in an unattended, automated way, providing at any time the right answers to the different users (from operators to engineers), with the man-machine interface appropriate for each one of them.

To add more complexity to the problem, the environmental constraints are becoming more and more complex and stringent. In the US, CO_2 emission reduction initiatives are a hot topic, like the State based emission reduction initiatives. For example, the California Climate Exchange Law AB-32, with a cap-and-trade scheme to be starting 2012 and the Regional Greenhouse Gas Initiative (RGGI), CO_2 Budget, whose trading program was put in effect in 2009. In the near future it is expected to have a US cap-and-trade policy for emissions reduction to be modeled on the EU scheme, that aims to cut carbon emissions to 1990 levels by 2020 and 80% by 2050.

There is a clear move towards a global carbon market. Carbon markets grew more than 3 times their worth in 2006 and it is expected to do so again by 2015, according to the World Bank. Companies have clearly identified that Carbon is money in the current and future economy while emission reporting is becoming mandatory worldwide. With all the regulations already on place and those coming on the horizon, emissions will become a significant factor in operations and planning, adding new operational constraints, increasing the operating cost, involving potential needs of capital investment and generating value from trading. In the path to compliance, organizations are facing several challenges in establishing an effective procedure to deal with the emissions under a consistent, optimum base.

And to add even more burden to the problem, control rooms worldwide are facing a shortage of people. Operators are concentrating more and more on the Units under his/her responsibility. DCS screens flourished and hundreds or thousands of tags are available per Unit and a multitude of process diagrams and trends can be projected on the walls and even the ceilings. But only a few employees are available to take care of the overall energy system operability and economics. Industrial energy systems are becoming increasingly complex and inter-related, not only between Units of the same Site but also with the power and utilities systems of neighbor facilities.

The evolution from having plant information scattered throughout many islands of automation to a unified and centralized Plant Information System was a clear enabling layer for such work. DCSs (Distributed Control Systems (DCS) and Plant information (PI) systems are the most widely available data sources in the current industrial context. PI systems usually acquire data from the DCSs and store it in a unique repository. The long term, facility wide PI system based historians constitute what is known as an *enabling technology*, because they are the cornerstone on top of which decision support systems, such as energy management systems, can be built. Centralized real-time databases provide access to massive current and historical process, laboratory and financial data.

Energy Management System Software Description and Functionality

In order to successfully address all the issues mentioned above, a software tool called Visual MESA has been extensively used. The original MESA program was developed in the early 1980's. Since the Visual MESA version was released in 1997, it has been continuously improved and adapted to be able to cope with all these changing scenarios.

Visual MESA is a computer program designed to model steam, boiler feed water (BFW), condensate, fuel, emissions and electrical systems. It is an on-line program that receives live plant data from the steam, fuel, condensate, BFW, and electrical system metering devices via a standard link to the plant information system.

It typically works uninterruptedly retrieving on-line data from the process using the OPC standard connectivity, evaluating continuously the utilities and energy system optimization alternatives and writing key performance indicators back to the plant information system.

The software includes all the necessary features to calculate and handle emissions properly: compositional description of the fuels streams, combustion properties to calculate boilers and heaters efficiencies, emission source stoichiometric calculations or estimated based on correlations for individual process point sources (for example, FCC catalyst Carbon burnt at the regenerator), heaters and boilers, etc.

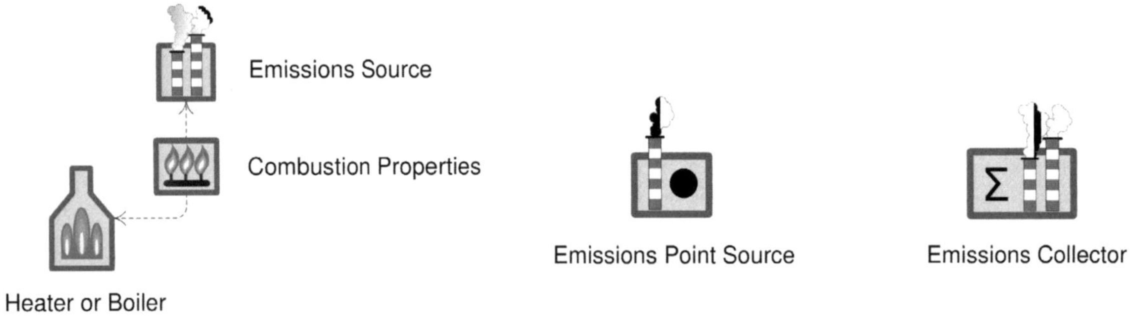

Figure 1 – Standard Visual MESA Emission Modeling blocks

Main benefit sources from where high savings can usually be obtained are:

Optimization: It helps to find the most economical way to run the utilities and energy systems, while remaining within the real operating constraints. It enables plant personnel to:

- Understand how to improve the optimum steam/fuels/electrical system operation, including emission constraints and costs.
- Quick reaction to minimize the economical impact of an operational change.
- Optimize the overall cost of fuel for steam generation and electricity, including the choice of the most convenient combination of turbines and motors.
- Make a good use of the utilities network data availability and improve its quality.

Monitoring: It provides a number of monitoring features that help access data, control data quality, and alert about changes to the system, like:

- Plant, equipments and streams information
- Trending data and calculation results

- Big changes on utilities streams and emissions regulations compliance alerts
- Data quality (balloons, graphically representing header mass balance imbalances)

Case Studies ("What If?" Planning): It allows performing and evaluating "What If?" cases that enable plant personnel to find ways to operate more efficiently and at less cost. Some examples are:

- Front-end loading on projects
- True "Plant-wide" project evaluation
- Steam, fuel, BFW, condensate, and electrical system improvements
- CO_2 emissions mitigation projects, credits availability and/or new constraints

Auditing & Accounting: It helps plant personnel to find where steam waste is occurring in the steam system, provides information to account for it properly and identifies where the imbalances are and how they behave in time.

This on-line energy management system was implemented in around 40 sites worldwide to model and optimize the overall site energy and utilities systems for a variety of industries: petrochemical, refining, chemical, sugar & alcohol, district heating and cooling, combined heat and power, etc. Several of them have been published (Refs. 4 to 17 are related to recent industrial projects). Amongst them, two particular examples will be presented and discussed.

Industrial Project Examples

The following two industrial projects are presented as examples:

- Medium size refinery complex
  - Example of daily recommendations and their impact on CO_2 emission
- Complex refinery with high conversion Units
  - Fuel gas network management and integration with the rest of the utilities system

Medium size refinery complex

This industrial example corresponds to TOTAL Feyzin refinery (Refs. 9 and 10). TOTAL is a leading refiner-marketer operating 11 refineries directly in Western Europe and one refinery in the U.S. at Port Arthur, TX. The reduction of greenhouse emissions and the enhancement of energy efficiency are among the current Company corporate challenges.

In this example, a set of manual operating recommendations given by the optimizer during a particular shift have been:

- Pump swaps (i.e., steam turbine and motors driving pumps that can be used interchangeably)
- Fuels to boilers, such as Fuel Gas (FG) and Fuel Oil (FO)

- Power production in steam turbine generators

As a result of the manual actions, adjustments to the following variables have been made by the control system:

- Steam production from boilers
- Letdown and vent rates

The following figures show the impact on steam production, fuel use and CO_2 emissions reduction:

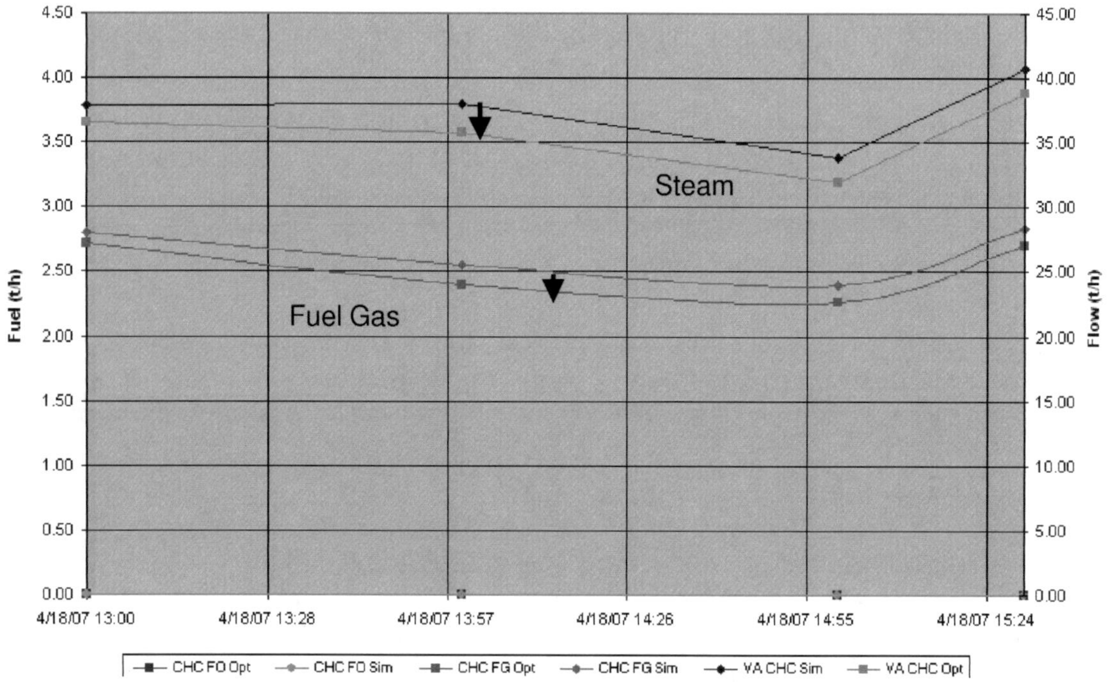

Figure 2 –Boiler C (100% Fuel Gas): 2 t/h less steam

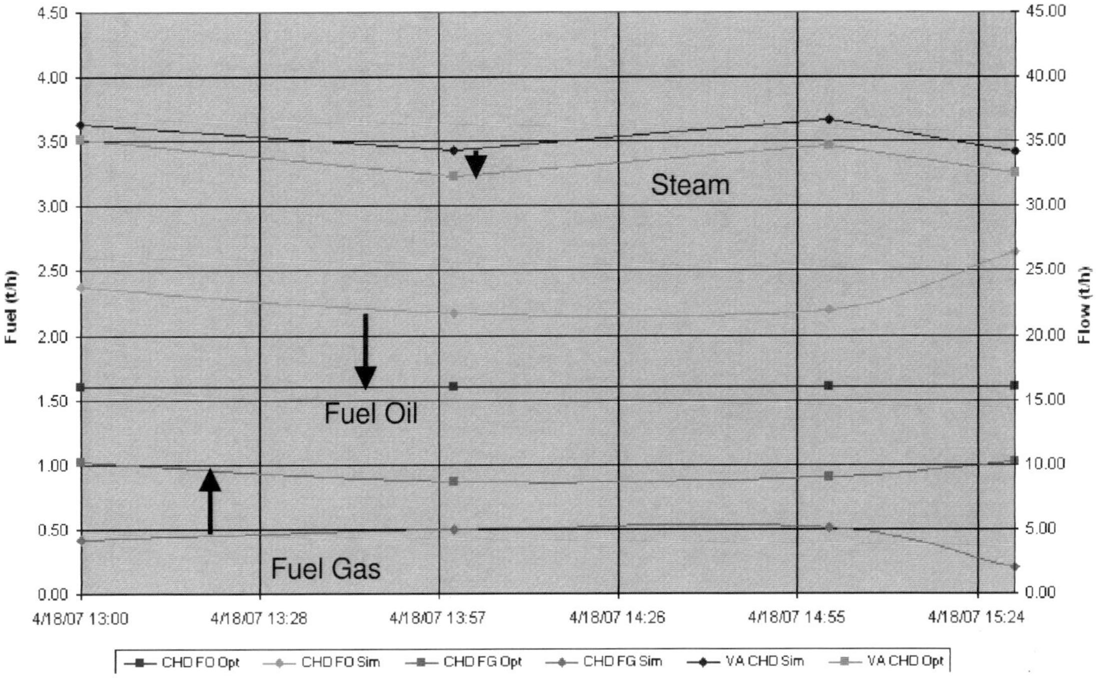

Figure 3 – Boiler D (FO and FG): 2 t/h less steam) and FO to the minimum

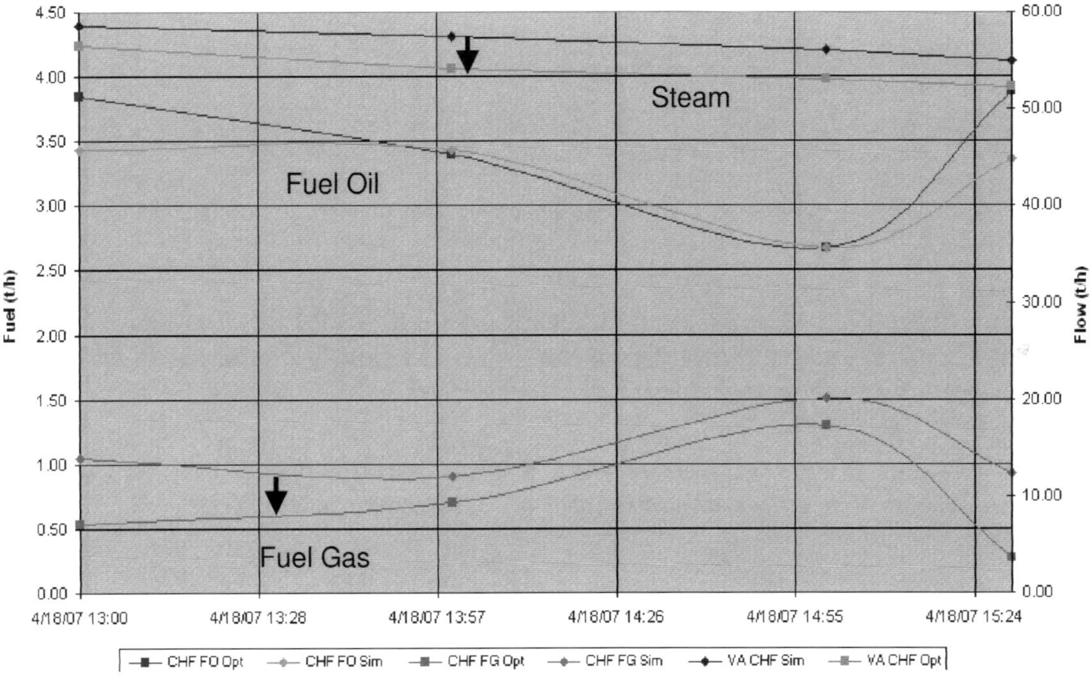

Figure 4 – Boiler F (FO and FG): more than 3 t/h less steam

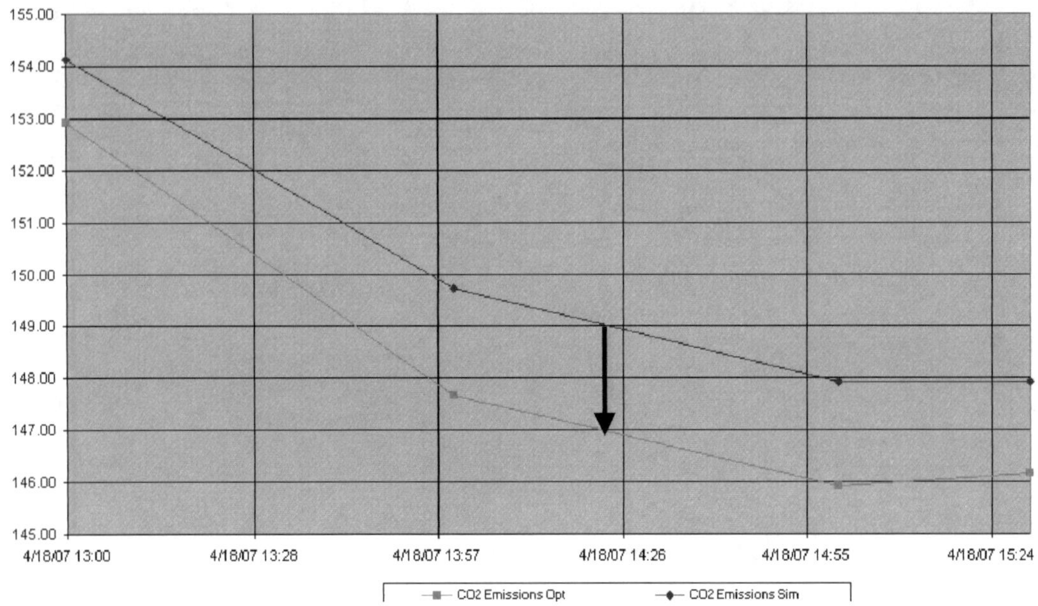

Figure 5 – CO_2 emissions reduction: 2 t/h less

In summary, the following changes were achieved while optimizing the overall cost:

- Almost 1 t/h less of FO consumed
- Approx. 7 t/h less of high pressure steam produced
- Approx. 2 t/h less of CO_2 emitted
- Approx. 200 kW more of electricity imported

When considering the CO_2 emission cost, the following is an example of the recommendation for operators:

As a result of the manual change in the boiler fuel use (more FG and less FO in overall), the Fuel Gas header pressure control system made the necessary adjustments which resulted in the need to increase the incremental Natural Gas import.

The following figure shows the fuel gas network model representation highlighting the differences between current and optimized situation (delta view, duty flows).

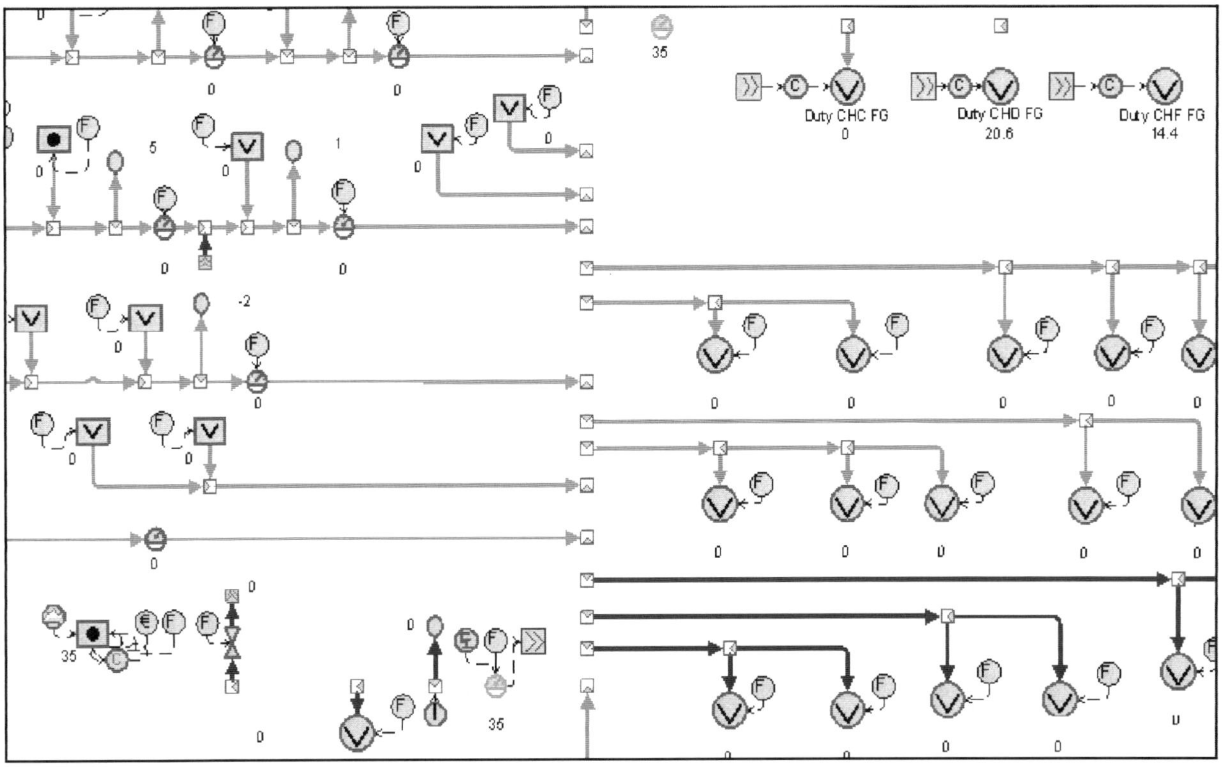

Figure 6 – Fuel Gas model showing a “delta view” (i.e., optimum – base case change)

On the left, fuel gas suppliers are represented, while on the right Fuel Gas consumers are displayed. The values indicate the corresponding changes, expressed in MW, after the application of the recommendations (zero value means no change).

As a result of replacing FO by FG (with the need to import more Natural Gas) and improving the steam efficiency usage by reducing let-downs and vents, CO_2 emissions are reduced by 4.7 t/h. This is important to consider when a trade-off between cheaper fuels that produce more CO_2 and more expensive fuels that produce less CO_2 exist. It would be an especially important consideration when the CO_2 emissions quota is expected to be exceeded by the end of the year.

Complex refinery with high conversion Units

At the Petronor site, Spain’s largest refinery (ref. 1) with a processing capacity of eleven million tons of crude oil per year, a Visual MESA based on-line model for energy management has been implemented. Its energy system is large and complex; therefore, the auditing and control of energy costs are a real challenge.

The full model of the energy system includes the entire fuel, steam, boiler feed water, condensate and power systems. Steam is generated in several places, with both conventional boilers, heat recovery steam generators and a cogeneration unit with steam injection. Electricity and fuel supply contract details have been included in the model. In particular, the purchased Natural Gas daily quota was rigorously modeled and the constraints respected during the routine optimization.

All the constraints have been included and the model is continually populated with live, validated data. Performance monitoring is performed and includes the follow up of calculated equipment efficiencies using the validated data mentioned before as well as the site Carbon emissions.

By auditing the energy system, imbalances are detected and diagnosed. As a result, the data can be relied on for evaluating the value of energy production and usage, and waste can be eliminated. Planning for a better operation of the energy system by performing case studies is usually done by using the validated model in off-line mode. Finally, it is important to mention that the optimization of operating conditions is performed on a continuous, daily basis since the system was commissioned in 2005. As a result of the project, substantial savings in energy costs have been achieved.

In order to illustrate the obtained benefits, optimization results are shown in Figure 7. Results correspond to one day of operation using Visual MESA. Savings are indicated as a percentage of total energy costs. After implementing the optimization recommendations, operation is closer to the minimum global energy cost, as it can be seen in the plot, after 9:36 AM. When the operator saw in the report potential savings higher than 1% of the overall utilities system cost, he implemented the optimization recommendations (in this case, a combination of changes in boiler loads and steam flows among production areas).

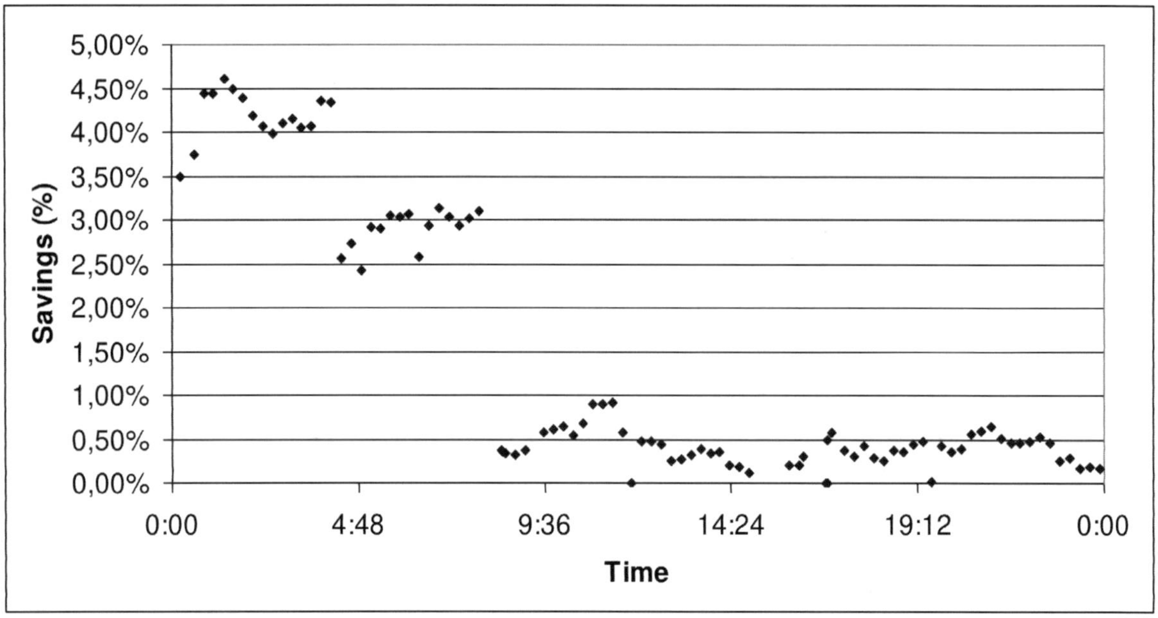

Figure 7 – Energy Costs Reduction Performance: One Day of Operation

Real Time Energy Management System Projects Key Issues

To successfully implement, deploy and sustain a real time energy management system energy cost and CO_2 emission reduction project, there are several issues that need to be addressed and recommended practices we have found useful to be taken into account.

Proper Technology Selection: Select the technology that fits with the problem solution rather than force the problem to be solved with an a priori selected technology. For example, if a simple linear regression with two input variables is sufficient enough to infer the NO_x emission, do not force the use of a several inputs neural network. But always high rank systems based on sound process engineering principles (i.e., mass and energy balances and rigorous non-linear equipment modeling) than those based on correlations or non-physical sense solutions (Friedman, 1999).

Avoid "Computerization": To solve a given problem, engineers need to apply the *EBC* approach suggested by Richalet, 2000, rather than the straight use of *C*. Follow the *EBC* procedure implies:

- First, use your ***E***ye to inspect trends, data, process diagrams, DCS screens...
- Second, use your ***B***rain to carefully understand and design the real time system and select or design the proper software tool.
- Finally, only at the end, use the ***C***omputer to solve the problem.

The enthusiastic but inexperienced engineers tend to go straight to the ***C***omputer, forgetting or minimizing to analyze and perfectly understand the problem to be solved, their real handles and constraints.

Balance Model Complexity against Expected Results: Energy system model complexity must be balanced against expected benefits. Remember the golden rule: "the simplest is generally the better". Engineering judgment and experience plays a key role in this task. Operators and Engineers will appreciate any efforts towards simplification, especially for future re-training and system maintenance.

Real Time Data Seldom Can be Used "as is": Carefully designed and engineered "safety envelops" must be built around real time on-line data, in order to validate all the signals, in and out of the applications. Minimum and maximum limits for each variable, controller windup identification and associated valve openings and controllers statuses must be combined to properly identify and isolate measurement problems. Steady state detection procedures must also be considered to warn Operators or directly skip an optimization run if the Process is unstable. What should a system do if bad data is detected? If the bad data feeds "critical" variables of the system, then its execution should be suspended and Operators warned. If bad data is related to "noncritical" variables then some kind of backup or default values could be used but such an action needs to be defined on each individual variable. The implementation of data validation procedures is a key activity for every real time energy management system project.

Pricing Information Must be Properly Updated: As economics is the main driver to apply real time energy management systems, prices used for the optimization or objective function calculations must be diligently updated. Manual data entering should be avoided.

Automatic pricing information must be implemented acquiring the prices directly from process information systems or web pages (some examples we have used in the past include the web sites of ERCOT in Texas, US, OMEL in Spain, both for electricity market spot prices and ICE in the US for energy prices).

Have Operators Always in Mind: From the very beginning try to involve Operators, inform them about the project objectives and solve their concerns. A common question is about being replaced by the new system and fired after completion. Second in rank is their worries regarding if the new system will increase their burden. First question is usually outside our control but we never have seen a single Operator being fired due to our projects, all of them successful, by the way. With respect to the second concern, an especial effort needs to be done to not increase Operator's burden, implementing automatic data updates or procedures to minimize his/her manual intervention. Clear and easy to understand operating interfaces need also to be built. Operators must be properly trained and assisted right by the DCS console after being left alone with the new system. It is always a good idea to spend many hours at the DCS console with the Operators to discuss the proposed solutions and technology, explaining to them what is needed so they will accept the introduction of the new technology. Properly designed Operating manuals must be provided and updated as frequently as needed.

Do not "Prototype" but Deploy Well Engineered Systems: Do not expose Operators nor final users to "interim or prototyping versions" but only to finished, very well built models, including the Operating interface and true handles limits and constraints. Operator's general attitude is to closely adhere to their first impression. There is a single "silver bullet" to use with them: they will never forget if early disappointed or frustrated when trying to use the system for the first time. Moreover, many of them would became "vaccinated" against the energy management technology and refuse to use it, subtle or frontally.

Sustain the Application to be Maintained "Forever Green": Engineering documentation must be provided and updated as soon as changes are implemented in the model or to the Operating interface. The application needs to be maintained "forever green", reflecting continuously the current Process handles and constraints. Changes on the energy system or process related side must be immediately reflected in the model. Local resources need to be properly trained and technology providers need to assist them and remotely maintain the applications. As a side note, it is important to mention the huge impact of remote access capabilities on systems performance. Only a few years ago just a minor model adjustment would take a lot of time for the new information to be exchanged back and forth. Nowadays, VPN or Webex like secure remote access can be easily implemented and solutions can be provided in minutes, instead of hours or even days.

Conclusions

CO_2 emissions and energy cost reductions have been obtained taking advantage of an energy management system software functionalities. It demonstrated to be a robust optimizer and very well suited to be used on-line, providing optimization recommendations to the operators on a routine basis.

The authors, working tightly with clients and Soteica engineering team members, following the generic outlined project rules, have participated in many successful implementations which resulted in significant economic benefits for the refining and petrochemical industries involved.

The success ratio of the industrial real time project deployments was highly increased by following a set of simple rules and ensuring a proper knowledge transfer to both, Operators and Engineers. Coordination among plant areas in order to implement the proposed optimization recommendations, as well as management involvement, were critical issues. The robustness of the tool helped Operators to gain confidence on the overall optimization system.

Application sustainability, which includes both maintenance and support, helps to sustain the economic benefits after commissioning allowing them to be always in good shape and still used and appreciated by the owners many years after handing the project over to them.

References

1. OPC Foundation, http://www.opcfoundation.org/, 2010
2. "Avoid Advanced Control Project Mistakes", Y. Z. Friedman, Chapter of the book Advanced Process Control and Information Systems for the Process Industries, L. Kane (editor) , ISBN 0-88415-239-1, Gulf Publishing Co., Houston, Texas, Pages 42-48, 1999
3. Personal communication, J. Richalet, 2000
4. "Auditing and control of energy costs in a large refinery by using an on line tool", D. Ruiz, C. Ruiz, J. Mamprin, Depto. de Energías y Efluentes Petronor, (2005), ERTC Asset Maximisation Conference organized by Global Technology Forum, Budapest, May 23-25, 2005
5. "Reducing refinery energy costs", Ruiz D., Ruiz C., Nelson D., Roseme G., Lázaro M. and Sartaguda M., (2006), Petroleum Technology Quarterly, pages 103-105, Q1 2006
6. "Online energy management", S. Benedicto, B. Garrote, D. Ruiz, J. Mamprin and C. Ruiz, Petroleum Technology Quarterly, pages 131-138, Q1 2007
7. "The Use of an On-line model for Energy Site-Wide Costs Minimisation", García Casas, J.M., Kihn, M., Ruiz D. and Ruiz C., (2007), ERTC Asset Maximisation Conference organized by Global Technology Forum, May 21-23, Rome, 2007
8. "Energy System Real Time Optimization", D. Uztürk, H. D. Franklin, J. M. Righi, A. T. Georgiou, NPRA Plant Automation and Decision Support Conference, Phoenix, USA, 2006

9. "Site-wide Energy Costs Reduction at TOTAL Feyzin Refinery", Département Procédés - Energie, Logistique, Utilités, TOTAL - Raffinerie de Feyzin, D. Ruiz, J. Mamprin and C. Ruiz, ERTC 12th Annual Meeting organized by Global Technology Forum, 19-21 November, Barcelona, 2007
10. “A Watchdog System for Energy Efficiency and CO2 Emissions Reduction”, Diego Ruiz, Carlos A. Ruiz, European Refining Technology Conference (ERTC) Sustainable Refining, Brussels, Belgium, April 2008
11. “Online energy costs optimizer at petrochemical plant”, Marcos Kihn, Diego Ruiz, Carlos Ruiz, Antonio García Nogales, Hydrocarbon Engineering, pages 119-123, May 2008
12. “Finding Benefits by Modeling and Optimizing Steam and Power Systems”, M. Reid, C. Harper, C. Hayes, Industrial Energy Technology Conference (IETC), New Orleans, May 2008
13. “Energy Real-Time Optimizer Baytown Refining/Chemical Complex”, M. Lorenz, Industrial Energy Technology Conference (IETC), New Orleans, May 2009
14. “Real Time Energy Management at Repsol Cartagena Refinery”, Fernán Serralunga, Diego Ruiz, Carlos Ruiz, Process Department REPSOL Cartagena Refinery, European Refining Technology Conference (ERTC) Asset Maximisation, Prague, May 2009
15. “Industrial Experience on the Implementation of Real Time On Line Energy Management Systems in Sugar and Alcohol Industry”, Douglas C. Mariani, Marcos A. Kihn, Carlos A. Ruiz, 10th International Symposium on Process Systems Engineering - PSE2009, Salvador de Bahia, Brazil, Elsevier, August 2009
16. “Emissions and Energy: An Integral Approach Using an On-Line Energy Management and Optimization Mode”l, Diego Ruiz, Fernán Serralunga, Carlos Ruiz, European Refining Technology Conference (ERTC) Annual Meeting Energy Workshop, Berlin, Germany, November 2009
17. “Online real-time optimisation helps identify energy gaps”, Marcelo Galleguillos, Michel Maffet, Marcos Kihn, Ruben Monje and Carlos Ruiz. Petroleum Technology Quarterly (PTQ), Q4, 101-109, October 2010

(51b) Reduction of CO_2 Emissions In a Refinery

Jason Stalhman, UOP, Des Plaines, IL

The CO2 Emissions from a refinery may be reduced in a variety of ways. This presentation will focus on methods to optimize Utilities and Offsites related equipment; both inside and outside the process unit battery limits. By reducing fuel fired in Utility Boilers, and CO2 production in other areas of the refinery, the total CO2 emission footprint will be reduced. Specific topics that will be discussed: optimization of an MP steam header pressure, steam generation, multiple BFW headers, HRSG supplemental firing, and tank vapor recovery.

See more of this Session: CO_2 Reduction

See more of this Group/Topical: Topical 7: 14th Topical on Refinery Processing

(51c) Reducing CO_2 Emissions In FCC Process

X. X. Frank Zhu, UOP LLC, Des Plaines, IL and **Saadet Ulas Acikgoz**, R&D Adsorption and Gas Processing, UOP LLC, Des Plaines, IL

The FCC process unit has been a major work horse in the petroleum industry since the 1940's. The attention for improving FCC performance in the past has been on enhancing the reaction conditions and catalysts to deliver value through yield improvements. However, in the current high energy cost environment and forthcoming CO_2 regulations, reducing CO_2 emissions in FCC unit started drawing much more attention. Refineries contribute to ~4% of total CO_2 emissions in the United States. FCC units account for 15-20% of refinery energy consumption and CO_2 emissions. In this project, novel and major improvement opportunities for FCC have been identified which can create a step change in reducing CO_2 emissions in FCC units.

In summary, improvement opportunities are identified based on viewing FCC as ISBL (Inside Battery Limit) and integrating FCC with the refinery as whole, i.e. OSBL (Outside Battery Limit). The opportunities include optimization of fractionation/separation flowsheet and conditions, integration of advanced technologies for fractionation/separation, maximizing heat and power recovery within FCC and integration of generated heat and power in ISBL with OSBL balances. Collectively, these improvements could help establish the image of FCC as a utility hub, delivering high value energy utilities to the refinery and lessen the CO_2 emissions from the refinery.

The new concepts developed as a result of this project include:

- *New flue gas cooler design:* Integrating a raw feed preheating system into the flue gas cooler to reduce flue gas cooler discharge temperature and maximize energy recovery.

- *Dividing wall column for naphtha splitting:* Using dividing wall columns for naphtha splitting and column sequencing to minimize utility usage and capital costs

- *Main fractionator optimization:* Improving the design of FCC Main Fractionator to recover more energy

- *Steam turbine integration:* Extraction HP (High Pressure) steam turbine in power recovery section to provide all the MP (Medium Pressure) and LP (Low Pressure) internal steam requirements in FCC unit and maximize power recovery

(51d) Profitable Emission Reduction In Petroleum Refineries

Jian Zhang and Qiang Xu, Dan F. Smith Department of Chemical Engineering, Lamar University, Beaumont, TX

Emission reduction is becoming increasingly important for petroleum refineries, as more environmental concerns and stringent regulations have been posed. Air emission reduction in refineries is also a very challenging task. It involves a large-scale complex manufacturing system. In such system, air emissions could spatially scatter at many reactors, furnaces, and flaring facilities; and temporally occurs at any time. Thus, cost-effective emission-reduction strategies should neither depend on end-of-pipe approaches, nor focus on solutions for single units or subsystems. The major thrusts should focus on the comprehensively study of material, energy, and information exchanges within the entire manufacturing system, so as to identify the best solutions to the entire plant. From the systems theory point of view, it's unnecessary to enforce emission sources from each individual facility to reach the minimum. As long as the generated emission sources from some facilities could be appropriately reused by others facilities to the maximum extent, the entire system emissions may still have chances to approach the minimum. Therefore, to optimally balance emission source generations and utilizations among the entire system might be the smartest way to reduce refinery emissions.

A greedy concept of profitable emission reduction (PER) is proposed in this paper, which has merits of economically attractive, environmentally benign, and technologically viable for air emission reductions in petroleum refineries. To identify PER strategies, a methodology framework and a new general plant-wide scheduling model have been developed, which couples generic production activities and the characterization of major air emissions from refineries, such as CO_2, VOC (volatile organic compounds), NO_X (nitrogen oxides), and PM (particulate matters). A case study has demonstrated the efficacy of the PER concept and the developed methodologies.

See more of this Session: CO_2 Reduction

See more of this Group/Topical: Topical 7: 14th Topical on Refinery Processing

(53a) Biogas Technology: The Energy, Environment and Health of the Developing Nations

Oluyemi Olasoji Adetule, Environmental Health, University of Ibadan, Nigeria, Ibadan, Oyo State, Nigeria

There are to date more than 3.5 billion people, mostly in developing countries, who still rely on coal and biomass—such as wood, dung, and crop residues—as their main source of energy for both cooking and heating. These traditional sources of energy can be extremely polluting and cause serious environmental health problems. Women and children are at greater risk because of household responsibilities and increased exposure to air pollution. Biogas technology through a proper anaerobic digestion process offers a way out as it produces a gas that can be used for cooking without any environmental or health problem. This paper described the management of Cattle wastes through anaerobic digestion (AD) technology which has not been explored as an alternative energy production, pollution reduction measure and in tackling health-related problems associated with cooking. To demonstrate the potential of this technology, a $5m^3$ underground digester was constructed and used for biodigestion of cattle dung . The study was based in Ibadan, in southwest Nigeria. The structures were built of concrete components. At mesophilic temperature range, the digester produced 1.5 - $2.5m^3$ of biogas which was purified by passing through columns containing iron nails, lime and calcium chloride. The anaerobic digestion of the organic wastes produced environmentally friendly and energy efficient gas.The gas produced little or no smoke compared with cooking with wood, dung or crop resideues.The user does not also complain about any health related problem. Nigerian Government is now working out a renewable energy policy which emphasizes renewable energy production in order to tackle environmental and health problems associated with the use of wood, dung and crop residues by households.

See more of this Session: Design of Sustainable Processes
See more of this Group/Topical: Environmental Division

(53b) Effects of Heavy Metals From Flue Gas on Algal Growth and Lipid Production and Their Distribution In the System

Katerine Napan[1], Reece Butler[2], Byard Wood[2], Ronald Sims[1] and Sridhar Viamajala[3], (1)Biological Engineering, Utah State University, Logan, UT, (2)Mechanical and Aerospace Engineering, Utah State University, Logan, UT, (3)Chemical & environmental engineering, University of Toledo, Toledo, OH

Flue gas from coal fired plants is a major source of carbon and heavy metal emission. Broadly, CO_2, SO_X, NO_X and trace metals are the constituents of flue gas which are of environmental concern. Capturing these airborne constituents using algae cultivation would allow initiation of a beneficial use program for flue gas and provide an environmentally sustainable alternative for coal fired plants. Biosorption properties, high growth rates, production on nonagricultural land, high lipid productivity of algae offer an encouraging alternative for heavy metal capture. Several studies have indicated the influence of heavy metals on algae. This study attempts to address the distribution and impact of heavy metals present in flue gas on microalgae production. Ten heavy metals namely, As, Cd, Cr, Co, Cu, Pb, Ni, Hg, Se, and Zn are being considered for this study. To understand the effects, algae strain (*Scenedesmus* sp.) is being grown in batch reactors using APS media with individual addition of the heavy metal at concentration likely to be present in the culture after permeate recycling. Total Suspended Solids and optical density are used as indicators for growth monitoring. Metal distribution and uptake are analyzed using ICP-MS. Lipid accumulation inside the algal cell is analyzed using gas chromatography.

Co, Pb and Zn at a concentration of 0.32, 1.09 and 8.8 mg/L were evaluated. The metal concentration in the permeate was close to zero after 18 days for Pb, whereas, Zn and Co were present in the permeate. Pb, Co, and Zn were found in the harvested algal biomass, mostly distributed in the intracellular portion than the extracellular portion of the cell. Pb, Co and Zn had been observed to adsorb to cell wall and reached a plateau within few days. Progressively these metals accumulated inside the cell. Pb, a not-essential nutrient, took longer to accumulate inside the cell wall, while Co and Zn, micronutrients, where uptaken since the beginning of the experiments. After 18 days of growth, around 25% of the Zn and Pb evaluated had accumulated mostly inside the cell while the Pb adsorbed to the cell wall was close to zero. Around 20% of the Co evaluated entered the cell while the Co adsorbed on the cell wall reduced slower probably due to dynamic equilibration with the Co available in the medium. Adsorption in the reactor walls was also observed. The growth and lipid production where practically not affected during the period of evaluation.

See more of this Session: Design of Sustainable Processes
See more of this Group/Topical: Environmental Division

(53c) Utilization of Municipal Wastewater for Cooling System In Thermoelectric Power Production

Iman Safari[1], Michael E. Walker[1], Javad Abbasian[1], Hamid Arastoopour[1], Ming-Kai Hsieh[2], David A. Dzombak[2] and David C. Miller[3], (1)Chemical and Biological Engineering, Illinois Institute of Technology, Chicago, IL, (2)Department of Civil and Environmental Engineering, Carnegie Mellon University, Pittsburgh, PA, (3)U.S. Department of Energy, National Energy Technology Laboratory, Morgantown, WV

The thermoelectric power industry in the US uses a large amount of fresh water. As the availability of freshwater becomes increasingly limited, alternative sources of water for power plant cooling are of interest for both existing and future power plants. Potential non-traditional water sources for use in thermoelectric power production include municipal wastewater, produced water from oil and gas extraction, mine drainage, and water produced from carbon sequestration. Utilization of impaired water requires evaluation of the deleterious effects on the cooling system and necessary treatments with consideration for quality and abundance of the impaired water source.

This paper addresses the development of a software module in Aspen plus® to identify and evaluate water management and treatment processes required for using municipal wastewater in thermoelectric power production inorder to reduce freshwater use. Establishment of the initial framework to model and simulate recirculating cooling system, treatment processes, and chemical equilibrium will be discussed. This framework will enable the end user to perform sizing, costing, and optimization analysis. The effect of different water properties on the selection and configuration of treatment processes, as well as performance of the main condenser and cooling water cycle, will be addressed. Finally, a costing analysis for two scenarios considering water of different qualities will be presented.

See more of this Session: Design of Sustainable Processes
See more of this Group/Topical: Environmental Division

(53d) A Graphical Approach to Flow Sheet Synthesis as Applied to the Fischer-Tropsch System

James A. Fox, Diane Hildebrant, Bilal Patel and David Glasser, University of the Witwatersrand, Center of Material and Process Synthesis, Johannesburg, South Africa

In this paper a novel approach to flow sheet synthesis is applied to the design of a flow sheet for the production of Fischer-Tropsch hydrocarbons. This graphical technique, called the GH-space, uses the flows of heat and work inside a process to provide insight into the interactions of various process units from the very beginning of a design.

Any unit process can be represented as a vector on the GH-space. By manipulating these vectors a process can be designed to meet certain design criteria or constraints.

In this paper five independent mass balances (reactions) were defined. These were then used; together with vectors describing mixing, separation and pressure; to determine the overall mass, energy and work balances for the process before any flow sheet existed. These overall balances were then used to begin formulating the flow sheet.

It was shown that with careful manipulation of design variables a process could be designed that not only produced the desired product but also consumed Carbon Dioxide as a feed and could potentially generate electricity.

While the GH-space technique is dynamic and flexible it tends towards the "best" thermodynamic solution. While it may not always be possible in a practical sense it allows a basis for more detailed design or for the modification of existing flow sheets.

See more of this Session: Design of Sustainable Processes
See more of this Group/Topical: Environmental Division

Reduce CO2 Emissions with ORC Low Heat Recovery

Dr. Ali K. Bourji, WorleyParsons, Houston, TX and Jimmy Winningham, WorleyParsons, Houston, TX

Organic Rankine Cycle (ORC) technology can recover waste heat from flue gas in onshore facilities and produce power. By use of select organic fluids, ORC makes significant heat recovery possible at lower flue gas temperatures than what is possible with typical Rankine cycles using water as the working fluid. With flue gas inlet temperatures as low as 290ºF, it is possible to incorporate ORC technology into plant designs or retrofits and recover electric power that can be used by the plant or sold to the power grid. By reducing power consumption, a facility can lower their CO2 emissions.

Figure 1 below, is a diagram which depicts an ORC. The cool working fluid is pumped to high pressure and sent to a heat exchanger where it is vaporized by cross exchange with the hot flue gas. The high pressure, vapor phase working fluid then enters an expander/generator package where by process of expansion, produces work in the form of electrical power. After expansion, the low pressure fluid is cooled and condensed in a condenser and the fluid returns to the inlet of the pump. The power generated in the expander is enough to power both the condenser and the pump and have a net output.

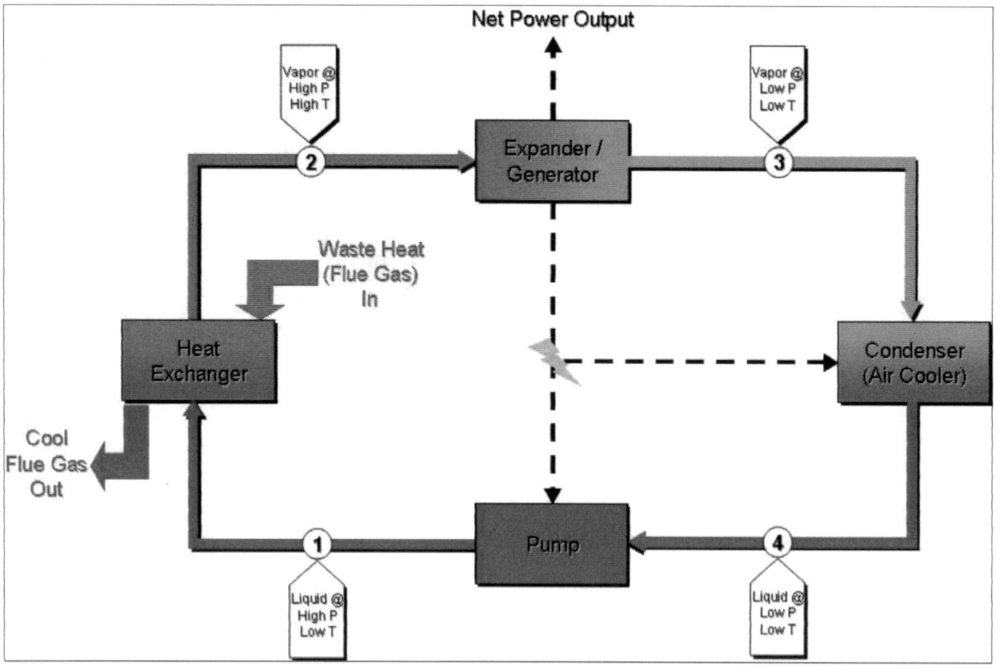

Figure 1. ORC System

Flue gas exits the ORC systems at temperatures low enough for introduction into flue gas treatment systems for CO2 recovery. Thus, ORC technology, installed in series with flue gas treatment systems, can recover a portion of the energy that must be removed from treated flue gas, rather than dissipating that energy as waste heat to cooling water or to the atmosphere.

A typical slate of flue gas from refinery fired heaters was used in a process model of an ORC system that uses refrigerant R-245fa [1] as the working fluid. Units included atmospheric distillation and vacuum units, hydrodesulfurization units, reformers, fluid catalytic crackers, delayed cokers, alkylation units, hydrodealkylation units, and utility boilers. Refinery sizes in the capacity range of 100,000 BPSD crude and greater were considered. Flue gas temperatures in the study ranged from 290°F to 500°F. The contribution by individual units to the total refinery waste heat recovery is shown in Figure 2.

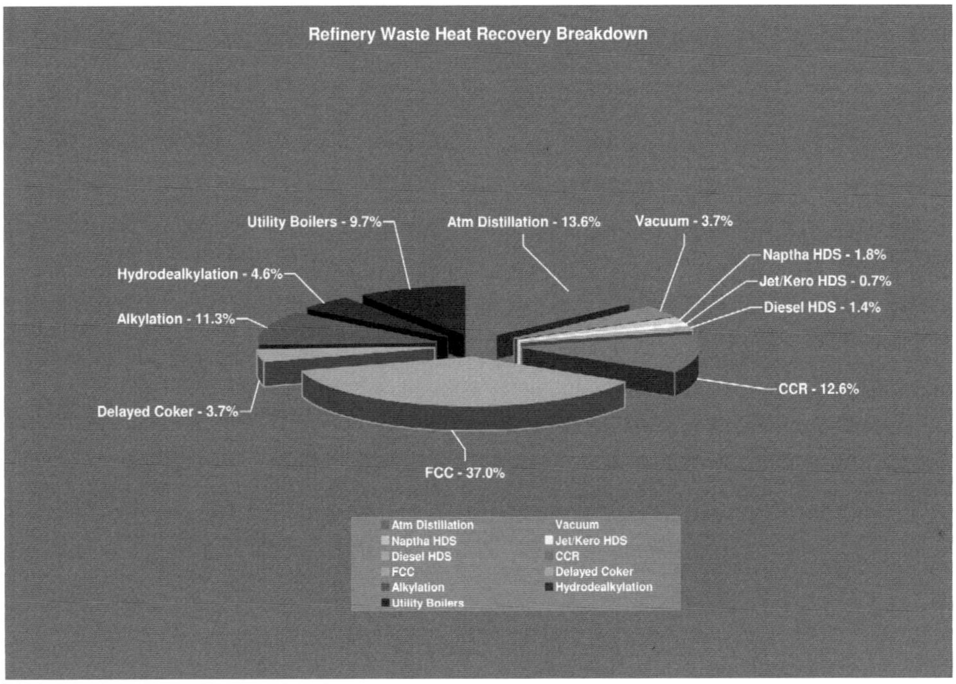

Figure 2. Refinery Waste Heat Recovery by Unit [2]

The estimated net electrical power an ORC generates from a typical flue gas in a refinery is shown in Figure 3. Results include power generation in kW-h per MMscfd of flue gas.

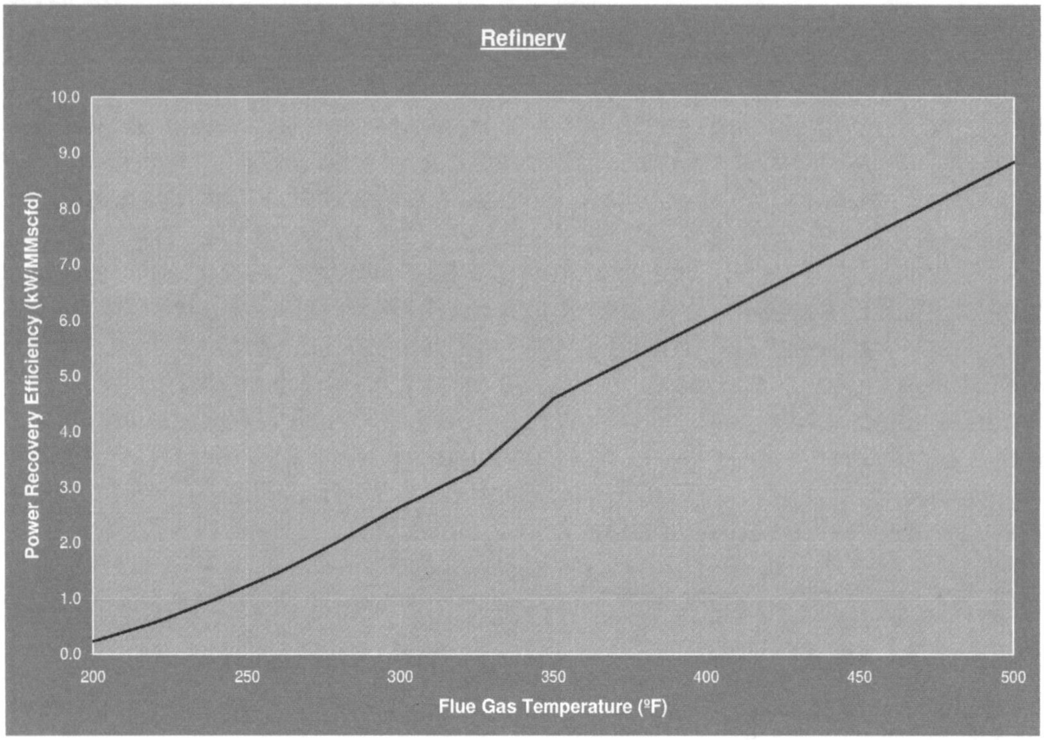

Figure 3. Recovery Efficiency versus Flue Gas Temperature [3]

The flue gas composition, temperature and flowrate used in calculations are shown in Table 1.

Table 1. Flue gas composition, temperature and flowrate used in modeling

Flowrate (MMSCFD)	Up to 1400
Temperature (°F)	290 – 500
Composition (MOL %)	
H2O	14
CO2	7
O2	3
N2	76
SO2	0

The process and economic data are fully scalable for plants, units, and flue gas flows and temperatures. The economics are further enhanced by including credits for CO2 emissions reduction.

The Lieberman-Warner Climate Security Act, S. 2191, has a direct effect on refiners that will "incur carbon costs for the carbon content of all of the refined products they produce and for emissions from fuels consumed in the process of refining… and modeling shows CO2 allowance prices for Lieberman-Warner (and the corresponding carbon tax) that rise from about $16 per metric ton in 2012 to about $65 per metric ton in 2030" as shown in the Table 2 below [4].

Table 2. CO2 allowance prices (measured in constant 2007 dollars)

Year	CO2 Price ($/metric ton)
2012	16
2015	21
2020	30
2025	44
2030	65

In December 2010, California, a leader to the rest of the country in environmental issues, passed a plan that places a cap on greenhouse gas emissions and establishes carbon allowances. The California Climate Action Registry (a voluntary greenhouse gas emission registry) does not have a protocol for energy efficiency, but there are international protocols for carbon credit generation for energy efficiency. Generally, in developed countries like the US, these credits would find their way into voluntary carbon markets. According to reports, "the cap-and-trade program is a key component in the overall strategy to reach the emissions-reductions goal spelled out in AB 32, the state's landmark climate-change legislation. The bill calls for greenhouse-gas emissions to be rolled back to 1990 levels by 2020, about a 15 percent reduction from today's levels." [5]

REFERENCES

1. Genetron® 245fa Applications Development Guide. Honeywell, 2000. (Genetron® is a registered trademark of Honeywell.)

2. Bourji, A., Barnhart, J., Winningham, J. and A. Winstead, "Recover waste heat from fluegas", Chemical Engineering, September 2010.

3. Bourji, A., Barnhart, J., Winningham, J. and A. Winstead, "Convert waste heat into eco-friendly energy", Hydrocarbon Processing, December 2010.

4. Harrison, D., "Impacts of Potential Climate Change Policies on the Refining and Petrochemical Sectors", prepared for NPRA, April 2008.

5. Herdt, T., "California Air Board Adopts Cap-and-Trade Program", www.downstreamtoday.com, December 20, 2010.

(53f) Retrofitting An Existing Methanol Product Process Using Fundamental Process Synthesis Tools

Leonard Doctor Maluleka, **Bilal Patel**, Diane Hildebrandt and David Glasser, School of Chemical and Metallurgical Engineering, University of the Witwatersrand, Johannesburg, South Africa

Fundamental tools to determine the most efficient structures of processes, optimization of process sections, reducing emissions and identify efficient ways to operating processes have been previously developed (Patel et al, 2007; Baraka et al, 2010). Retrofitting is a term used in chemical engineering to define updating the outdated. Retrofitting is applied on the improvement of flow-sheets, process equipment and chemical processes.

This investigation evaluates ways in which these tools can be utilized to analyze and retrofit an existing methanol production process. The process synthesis tool utilizes physical and thermodynamic principles to set targets. The targets provide insight to the mass, energy and entropy flows of the methanol production process. These targets can be compared to the actual flows, thus allowing for opportunities for improve process efficiency and providing insight for process integration.

Initially, we consider the overall mass balance for the process. The mass balance gives one a good indication of the overall carbon efficiency of the process. This overall mass balance is fixed, and also sets the heat and work requirements of the overall process. We can compare this overall mass balance, with the target mass balance. The next step is to look at the process sections i.e. syngas production and synthesis section. The amount of heat and work that is required by the sections is governed by the thermodynamic performance of these sections and the process integration between both sections. Again, these can be compared to the target values. This allows one to determine how efficient the process is and also determine the amount of lost work from each section. One can also compare the utility requirements for the process

The mass, heat and work flows in a process can be a complicated and often difficult to untangle the effects of these flows on the process efficiency. Using the novel tools, makes process analysis easier and more effective in making decisions in the early stages of design and also in retrofitting existing chemical processes.

See more of this Session: Design of Sustainable Processes
See more of this Group/Topical: Environmental Division

(53g) New Structure Design of Improved Zinc Oxide Nanowire-Based Solar Cells

Mallarie D. McCune, Chemical and Biomolecular Engineering, Georgia Institute of Technology, Atlanta, GA

NEW STRUCTURE DESIGN OF IMPROVED

ZINC OXIDE NANOWIRE-BASED SOLAR CELLS

Mallarie D. McCune- Ph.D. Candidate

Dr. Yulin Deng- Advisor

Georgia Institute of Technology, Department of Chemical and Biomolecular Engineering

-

Abstract

-

Because of its excellent and unique physical properties, ZnO nanowires have been widely used in numerous scientific fields such as sensor, solar cell, nanogenerators, etc. Although it is believed that single crystal ZnO has a much higher electron transfer rate than TiO_2, it was found that ZnO nanowire-based dye-sensitized solar cells have lower efficiency than TiO_2 nanoparticle-based solar cells, with the highest efficiency of 4.1%. To improve the ZnO nanowire based solar cell efficiency, different solar cell structures were proposed and related mechanisms were studied. The first approach was based on the addition of a block layer that will prevent electron back transfer at the anode, thus eliminating short-circuiting of the cell and increasing the cell efficiency. The second approach focused on the formation of a 3-D transparent conducting oxide network that enhances the electron transport within the solar cell. The combination of TiO_2 with ZnO nanowires for dye-sensitized solar cell application was also investigated. All of my research objectives are vital to the universal research efforts dedicated to resolving the global energy crisis.

(67a) Agency - Friend or Foe?

Anju Bhatia, Bureau of Air; Field Operation Section, Illinois Environmental Protection Agency, Des Plaines, IL

Permitting a new plant, process, or modification can be daunting task. Early presentation and meetings with the Agency can help you climb the paperwork hill. In this session, a brief description of the various Bureaus and how they interact will be presented. Further information will be provided on the Bureau of Air which has the most complex rules and regulations. Field Operations section performs the enforcement on behalf of the State and the Federal government. In addition to their regular duties, they carry on the special requests for inspections for the Citizens of Illinois as well as the USEPA. Non-compliance issues will be discussed as to how they are evaluated and depending on the seriousness, how they are ultimately resolved. The permitting process will be discussed as to how to tackle a new source, source modification as well as process modifications.

See more of this Session: YP Tutorial: Environmental Topics On Regulation and Innovation
See more of this Group/Topical: Environmental Division

(67b) Chemical Engineering Careers (focus on environmental consulting)

Kurt Rindfusz, Eastern Research Group, Inc., Chantilly, VA

The profession of chemical engineering offers multiple, distinct career paths. Although many young professionals entering the professional world will spend the majority of their time in an industrial setting, working in important careers such as process engineering; many alternatives present themselves. This session will provide a brief overview of the variety of "specialties" our profession offers, including careers in: process engineering, academia, law, and government. This overview will be followed by a detailed discussion of a career path in environmental consulting. The speaker holds MS and BS chemical engineering degrees and will discuss his thought process regarding the decision of which path to follow after graduation and the ultimate decision to enter the environmental consulting world, followed by an overview of experiences from nearly 20 years on the job. The intent is to share real-life observations, share the pros and cons associated with various career choices, and enter into a dialogue (time permitting) with the audience.

See more of this Session: YP Tutorial: Environmental Topics On Regulation and Innovation
See more of this Group/Topical: Environmental Division

(67d) Environmental and Administrative Law 101 with Greenhouse Gas and Sustainability Update

Mary Ellen Ternes, McAfee & Taft, Oklahoma City, OK

This tutorial provides a short course for chemical engineers regarding the basic concepts and structure of administrative and environmental law. The course begins with a history of legal development, from English common law to the state of common law and legislation in the United States today. The course reviews basic concepts of administrative law, and outlines the mechanisms by which environmental regulatory agencies implement delegated authority, including delegation of power, appropriate exercise of discretion, rulemaking, investigative and penalty authority, adjudication and judicial review. Building upon these agency powers of implementation, the course reviews all major federal environmental statutes that are implemented by environmental regulatory agencies, and outlines their scope, goals and primary mechanisms for enforcement. The course then winds up with a summary of common law claims typically associated with environmental issues, and an update on the most controversial environmental legal developments over the past year.

See more of this Session: YP Tutorial: Environmental Topics On Regulation and Innovation
See more of this Group/Topical: Environmental Division

(89a) Phase Equilibria of Organic Sulfur Compounds for the Development of Sulfur Removal Processes

Erlin Sapei, Juha-Pekka Pokki and Petri Uusi-Kyyny, Department of Biotechnology and Chemical Technology, Aalto University School of Science and Technology, Espoo, Finland

Environmental concerns have led many countries worldwide to apply stricter regulations on sulfur levels in liquid fuels. These regulations place many challenges for the refining industries. Separation process design to accomplish the removal of sulfur compounds requires knowledge of vapor-liquid equilibria (VLE) of sulfur compounds with hydrocarbons, particularly their activity coefficients at infinite dilution. However, the investigations concerning the systems containing sulfur compounds are scarce. Therefore, more measurements are needed to develop the estimation methods and the thermodynamics models in the dilute systems.

The VLE of sulfur compounds and hydrocarbons published in the open literature have been reviewed and updated with our measurements. We have determined over 100 VLE data sets of systems containing sulfur compounds found in petroleum refinery streams (mercaptans, sulfides, disulfides, thiophenes, and tetrahydrothiophene) in different hydrocarbons (alkanes, alkenes, cyclic, aromatic, and oxygenated components) with a variety of apparatuses (recirculation still, comparative ebulliometer, dilutor, and static total pressure). Measurements for determining phase equilibria of organic sulfur compounds will be continued and more data will be published in the near future. The challenges of the VLE measurements for systems containing sulfur compounds are discussed. The simulation of the behavior of diethyl sulfide in hydrocarbon distillation is presented as a case study. The case study is based on the experimental results from laboratory scale 15 stage Oldershaw type column of Ikari A, Hatate Y and Maruyama A, (1985), Chem. Eng. Comm., vol 34, p. 241-251. The parameters of the Wilson activity coefficient model were optimized based on our measured data and previously published literature data. The effect of Murphree vapour phase efficiencies on the sensitivity of the simulation is discussed.

See more of this Session: Environmental and Water Solutions
See more of this Group/Topical: Topical 7: 14th Topical on Refinery Processing

(89b) Water Consumption, Characterization, and Conservation In the Refining Industry

Jason L. Stahlman, UOP, Des Plaines, IL

Water is used extensively in a refinery, as a liquid or as steam. Historically, water has been used liberally, with water minimization primarily occurring only indirectly, usually as a result of direct efforts to reduce energy consumption. With water availability becoming more limited in many parts of the world, reduction of fresh water usage in a refinery will become increasingly important.

This presentation will look at the many uses of water in a refinery and at water's many forms. It will look at water used in the process units as well as water used in the Utility and Offsites units. The presentation will look at methods and techniques for reducing water consumption in general and fresh water consumption in particular. Techniques examined will include ruse opportunities, energy reduction, and water treatment methods. The presentation will review water balances for recent jobs where water limitations were mandated. It will discuss these limits and other reasons why it may be desirable to reduce water usage.

See more of this Session: Environmental and Water Solutions
See more of this Group/Topical: Topical 7: 14th Topical on Refinery Processing

(89c) Water Reuse In the HPI: Challenges and Opportunities

Steve Bakas, Water & Process Technologies, GE Power & Water, Trevose, PA

A more complete abstract will be added later. The paper will discuss the technologies used to facilitate the reuse of water in Hydrocarbon process Industries.

See more of this Session: Environmental and Water Solutions
See more of this Group/Topical: Topical 7: 14th Topical on Refinery Processing

Refinery Wastewater Treatment using Heavy Canadian Crude Feed Stock

Rafique Janjua Fluor corporation, Sugar Land, Texas

AIChE Spring 2011

Key words: Tar Sands, SAGD, Desalter waste, Desalter Process, Desalter mudwash waste COD/BOD, Dissolved Oxygen, Mixed Liquor Suspended Solids, Food to Mass Ratio, Emulsion, Colloidal Silica & Clay, Fluor technology.

Extended Abstract: AIChE Spring 2011

Refineries using Canadian Crude as feed stock are experiencing high emulsion wastewater from Desalter mud wash underflow operation. This wastewater is high in COD/BOD loading which is exerting high loading to the already struggling aeration system, resulting in inadequate operation to meet the NPDES limits. Emulsion also plugs the aeration diffusers and further adversely impact the Oxygen Transfer rate and mixing of mixed liquor suspended solids. Desalter underflow/mud wash can be successfully treated to mitigate the shock loading to existing wastewater plants.

Introduction:

Heavy crude produced by SAGD process in upstreams operation, entrains large amount of colloidal silica, clay and heavy oil emulsion part it ends up in the crude sold to the down-stream processing refineries. Typical waste streams generated in the refineries is listed in the table below.

Table: 1

Sources of Refinery Wastewater Streams:

Utility wastewater;

- Cooling Tower Lowdown
- Boiler Blowdown
- Once Through Service Water
- Pump Seals water
- Water Treatment, Demin. Regeneration & Filter Backwash etc
- Contaminated Storm Water

Process wastewater:

- Crude Desalting
- Thermal Cracking/Visbreaking
- Distillation
- Fluid catalytic Cracking
- Catalytic Reforming
- Alkylation
- Catalytic Hydro cracking
- Isomerization
- Merox Process
- Sulfur Removal/Clauss Process
- Others

Table 1

Estimated Wastewater Flow Rates from Petroleum Refineries Processing Units

Process unit	Possible Wastewater Description	Approximate Wastewater Flow Rate (Gallons/Barrel of Crude Petroleum)
Crude	Salts, metals, suspended solids, hydrogen sulfide, ammonia and phenols.	2 - 3
Distillation Desalting	Hydrogen sulfide, ammonia, suspended solids, chlorides, mercaptans and phenols.	22 - 26
Fluid Catalytic Cracking	Hydrogen sulfide, ammonia, suspended solids, oil, phenols and cyanides.	13 - 16
Catalytic Reforming	Hydrogen sulfides, ammonia, suspended solids, mercaptans and oil.	5 - 7
Alkylation	Spent potassium hydroxide stream (hydrofluoric acid)	2 - 4
Thermal Cracking / Visbreaking	Hydrogen sulfide, ammonia, suspended solids, dissolved solids and phenols.	2 - 4
Catalytic Hydro cracking	Hydrogen sulfides, ammonia and suspended solids	2 - 3
Cocking	Hydrogen sulfides, ammonia, and suspended solids	1 - 2
Isomerization	Hydrogen sulfide, ammonia, caustic wash water and chloride salts.	1 - 2
Catalytic Hydrotreating	Hydrogen sulfides, ammonia, suspended solids and phenols.	1 - 2

Sulfur Removal / Claus Process	Hydrogen sulfide and ammonia	<1.0
Lubricating Oil Manufacture	Oils solvents and propane.	<1.0
Sweetening / Merox Process	Spent caustic, sulfides and ammonia.	< 0.1

(Data adopted from EPA publications)

Refinery Process Flow Diagram

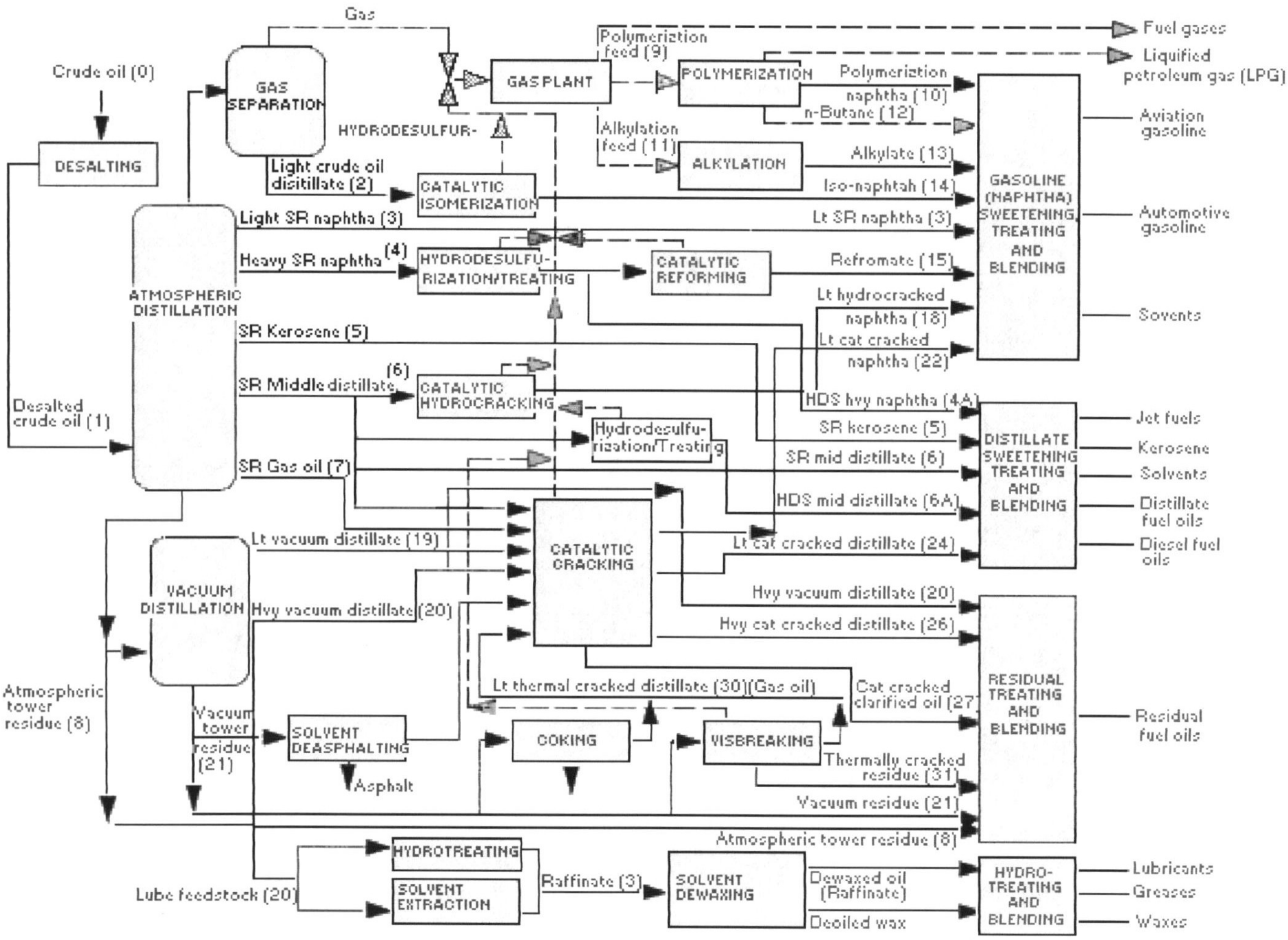

Desalting: (Crude Oil Pre-Treatment)

Crude oil often contains water, inorganic salts, suspended solids, and water-soluble trace metals. As a first step in the refining process, to reduce corrosion,

plugging, and fouling of downstream equipment and to prevent poisoning the catalysts in downstream processing units. These contaminants must be removed by desalting (dehydration).
The two most typical methods of crude-oil desalting, chemical and electrostatic separation use hot water as the extraction agent. In chemical desalting, water and chemical surfactant (demulsifies) are added to the crude, heated so that salts and other impurities dissolve into the water or attach to the water, and then held in the desalter vessel where they settle out. Electrical desalting is the application of high-voltage electrostatic charges to concentrate suspended water globules in the bottom of the desalter vessel. Surfactants are added only when the crude has a large amount of suspended solids. Both methods of desalting are continuous. A third and less-common process involves filtering heated crude using diatomaceous earth.

The feedstock crude oil is heated to between 150° and 350°F to reduce viscosity and surface tension for easier mixing and separation of the water. The temperature is limited by the vapor pressure of the crude-oil feedstock. In both methods other chemicals may be added. Ammonia is often used to reduce corrosion. Caustic or acid may be added to adjust the pH of the water wash. Wastewater and contaminants are discharged from the bottom of the desalter vessel to the wastewater treatment facility. The desalted crude is continuously drawn from the top of the desalter vessel and sent to the crude distillation (fractionating) tower.

In refinery operations, this step is called as "mud washing" and may take place more than once a day. Every mudwash operation may produce wastewater up to approximately 5 - 6% of the daily refinery crude throughput and can contain Oil & Grease, Salts, Metals, Suspended Solids, Hydrogen Sulfide, Ammonia, Phenols and Colloidal Silica and Clay emulsions.

In case of Canadian Crude feed stock it also contains a large amount of emulsified Colloidal Silica and Clay where COD can reach as high as 29,000 PPM and looks like as seen in the next picture.

This untreated waste stream when combines with rest of the refinery waste, plugs up the wastewater pre -treatment units such as API/CPI /DAF oil water separators and ends up in the bioreactor (aeration system) and it adversely impacts the biomass equilibrium. The first sign usually is the Dissolved oxygen level in the aeration basin dips to zero, mixed liquor suspended solids are suddenly increased, there's sudden increase in COD/BOD loading and food to mass ratio is changed. The operator has very little control to effectively respond to this overwhelming shock loading.

Why does this kind of waste come from Canadian Tar Sands Feed stock?

The answer lies in how the crude from Tar Sands deposit is extracted. One of the most common method called as Steam Assisted Gravity Drain (SAGD), is in situ heating of the tar deposit below the ground surface. Steam is the medium for heating the tar deposits. It is delivered at 1500 psig pressure (600F/315C) through horizontal drilled parallel pipes. Steam is injected from the top pipe and hot mixture of melted bitumen, condensate and dirt (containing dissolved and suspended matter) drains by gravity into the bottom pipe. The mixture is pumped out above ground where condensate (produced water) crude/diluted bitumen and solids are separated. The produced water is further treated for boiler feed quality for steam regeneration. There are a lot of challenges in how best to treat the produced water.

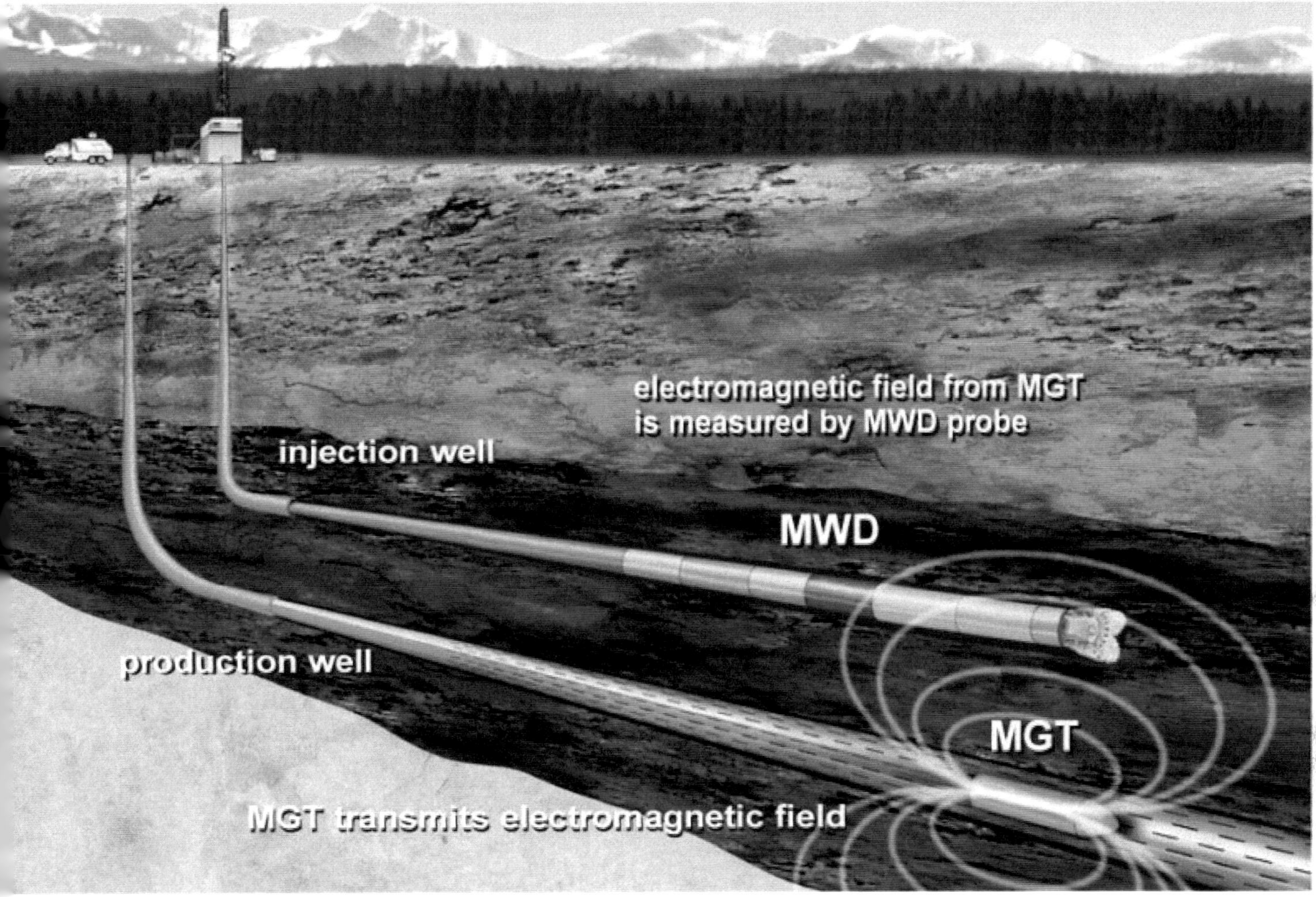

<u>**VIEW OF SAGD INSTALLATION**</u>

Produced Water Quality:
Ca = 35 ppm as CaCO3
Mg = 12
Na = 2638
K = 36
Fe = 0.8
HCO3 =76
Cl = 2604
SO4 = 31
SiO2 = 217 as SiO2
pH = 9+

Heavy emulsion of oil, grease, silica & clay. Some of this emulsion is carried in the crude oil and removed in the desalter underflow / mudwash waste.

Present Technologies to treat this kind of waste:

- Multiphase Centrifuges
  (CAPX/OPX)
- Physical-Chemical Treatment
  (acid/caustic feed, exotic metallurgy, high OPX & safety)
- V-Sep Membrane Technology
  (CAPX/OPX, several modular units)
- Others?

New and Innovative Technology by Fluor Corporation;

Selective chemical treatment can break the emulsion in the desalter underflow as seen here;

Jar Test Results

The precipitate is fairly sticky and fragile in nature, and dewaters quickly. The regular clarifiers using mechanical skimming devices are not effective in removing this layer.

The following sketch illustrates the FLUOR skim tank which utilizes no moving parts; it cumulates, thickens and hydraulically moves the precipitates over the fixed weir into the sludge tank for further treatment. Treated water is recycled to enhance the solids contact effect for optimum chemical reaction and rise rate variation. Sludge volumes and operating liquid level can be adjusted with out ever getting inside the tank by commercially available Guided Wave Radar instrument mounted outside the tank.

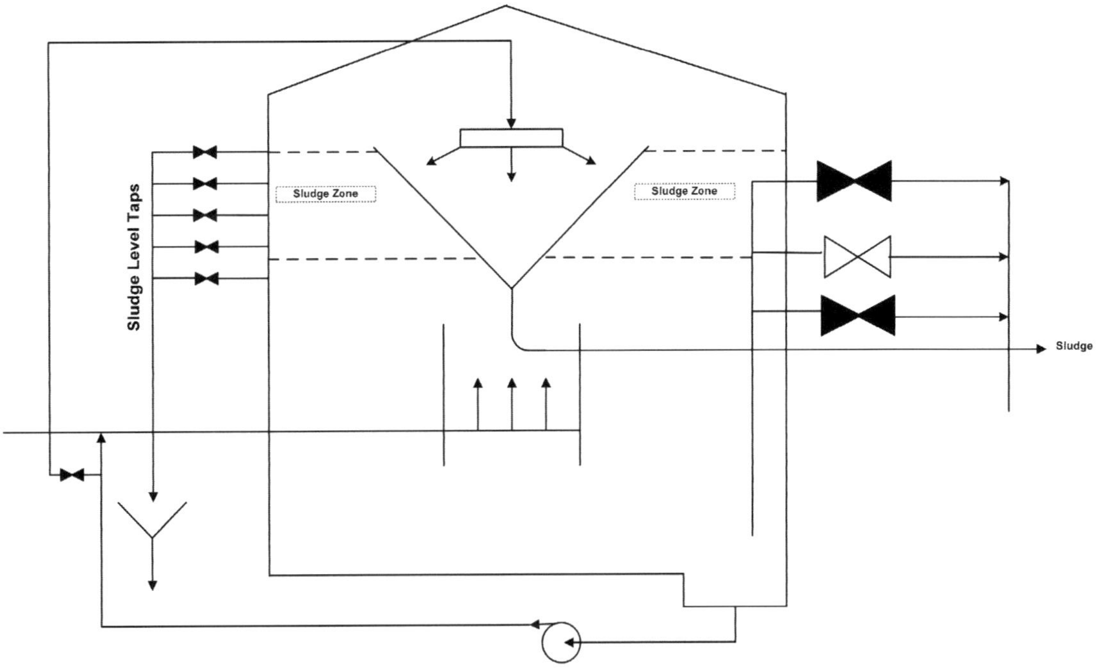

Separation Tank

Conclusion:

Refineries using Canadian Crude feed stock are experiencing a lot of operating difficulties in treating and producing effluent to meet the NPDES limits. This problem can be solved by segregating the desalter mud wash waste, using Fluor's technology by chemically treating to break the emulsion and skimming the floating precipitate. This process will reduce substantially the capital cost, operating cost and lower the COD/BOD loading (organic load to the bioreactor) to the level that the existing wastewater can process and produce acceptable quality effluent without major infrastructure up grades.

(89e) Challenges and Opportunities of Electrocoagulation for Refinery Application: Case Study with Chicken Processing Plant Water and Fleet Wash Water

Eric Peterson[1], Jewel Andrew Gomes[2], David L. Cocke[2], Hector Moreno[3], Morgan Reed[4], Joe W. Hutchins[5], Daniel Atambo[6] and Kamol Kanti Das[7], (1)Scandpower Inc., a member of the Lloyd's Register Group, Houston, TX, (2)Dan F. Smith Department of Chemical Engineering, Lamar University, Beaumont, TX, (3)Chemical Engineering, Instituto Tecnologico de la Laguna, Torreon, Mexico, (4) Drilling Fluids, Baker Hughes, Broussard, LA, (5)Process Improvement and Optimization, Lake Charles, LA, (6)Dallas/Fort Worth Regional Office, Texas Commission on Environmental Quality, Fort Worth, TX, (7)Agronomy, Purdue University, West Lafayette, IN

Due to the limited clean water resources of our planet, an urgent need exists for efficient technological development to treat wastewater for re-use. Several water treatment methods and instruments are currently available that partially achieve this goal. The cost and treatment efficiencies depend on the instrument type used and their application. Of these techniques, electrochemical methods, such as electrocoagulation (EC), can still provide the most efficient and economic service. EC works through in-situ formation of coagulants when anodes (e.g., iron, aluminum, copper, or titanium) are sacrificed electrochemically. Generally, direct current is applied through the EC cell. Research has been performed over decades using this process to successfully treat multiple types of wastewater. In this presentation, our experimental results on EC-treatment of chicken processing plant water and fleet wash water demonstrate how this may be accomplished. The floc produced during treatment was also examined using common materials characterization techniques (e.g., FTIR, XRD, and SEM-EDS). This work also illuminates challenges and opportunities necessary to make this technique viable for industrial application.

See more of this Session: Environmental and Water Solutions
See more of this Group/Topical: Topical 7: 14th Topical on Refinery Processing

(97a) Climate Change Legislation and Regulation

Mary Ellen Ternes, McAfee & Taft, Oklahoma City, OK

This presentation will constitute the majority of this session, and be presented by invited policy and legal experts. These experts will provide a review and update on the state, national and international legal and regulatory developments regarding greenhouse gas emission voluntary reduction and mandatory control programs. Specifically, this session will review the current and potential future Climate Change Legislation Landscape, including the structure of possible carbon constraints and implementation strategies, in addition to related issues including state and regional program preemption, California programs, EPA's regulation of Green House Gases (GHG) under various provisions of the Clean Air Act, and also current international GHG developments impacting US companies doing business overseas. This session will be of particular interest to those interested in greenhouse gas compliance strategies and approaches to new and development legislation and regulation.

See more of this Session: Legislation and Regulation of Greenhouse Gases and Chemical Products
See more of this Group/Topical: Environmental Division

(97b) GHG Measurement and Legal Implications

Mary Ellen Ternes, McAfee & Taft, Oklahoma City, OK and Chuck Hookham, HDR, A, CA

In the context of the Supreme Court's 2007 interpretation of the Clean Air Act in Massachusetts v. EPA, the EPA's subsequent GHG emission rulemaking, administrative challenges to this rulemaking and recent GHG litigation, this presentation will review current accepted GHG emission measurement methods, uncertainties in implementation and potential legal implications arising from this scenario. This presentation will compare and contrast different GHG measurement scenarios arising in varying contexts.

See more of this Session: Legislation and Regulation of Greenhouse Gases and Chemical Products
See more of this Group/Topical: Environmental Division

(97c) Challenges of Evaluating Regulatory Compliance with the GHG Mandatory Reporting Rule

Paul Glaves and Jane Buchanan, Advanced Applications - Environmental Practice, Mustang Engineering, Houston, TX

Compliance with the "Mandatory Reporting of Greenhouse Gases: Petroleum and Natural Gas Systems" for the gas processing and gathering industry has been challenging in many ways. Not only has the rule implementation timeframe been aggressive, but these new rules apply to a large number of facilities and sources that have not previously been required to be tracked and quantified.

This presentation will investigate the approach that was taken to identify the requirements and achieve regulatory compliance at major oil and gas operating company with Subparts C and W of the rule. The challenges include:

- Legacy facilities – what measurements are available and what meets the rule?
- Facility Definition Conflicts – Subparts C and W
- Measurements and Tracking
  - Rod Packing Venting Rates
- Compressor and station blowdowns as well as starter gas using natural gas
- Flaring events
- Compressor station treating and glycol dehydrator vents
- Data Collection
- Data Retention

The solutions adopted to address compliance will be discussed.

Global Implications of REACH and TSCA
Moving Towards Comprehensive Chemical Management Accountability

Jacqueline Sibblies, PE
Environmental Compliance Consulting
P.O. Box 5013, Somerset, NJ 08873
sibbliesj@gmail.com
(908) 531-4029

Introduction

The recent publication of the United States Environmental Protection Agency's (USEPA's) proposed rule to modify the Inventory Update Reporting (IUR) Rule under the Toxic Substances Control Act (TSCA) will certainly cause members of the chemical manufacturing and distribution industry to evaluate the impact(s) of any resulting new and/or modified regulatory requirements that could directly affect their businesses. And with many companies having international affiliations, the onset of changes in regulatory requirements in one global region would most likely affect operations in several other regions. Methods of manufacturing in one country, for example, may have to be revised in order to meet the product quality standards of another.

Then there is also the consideration that must be given to the effect of interrelated regulatory requirements. New or modified regulations could be burdensome to some members of the regulated community, while others may find the new requirements complimentary with existing subject regulations. For some chemical companies, this may be the case with the TSCA IUR and REACH (the Registration, Evaluation, Authorization of Chemical Substances regulations) administered by the European Chemical Agency (ECHA). This paper will present a summary of key requirements of both REACH and TSCA IUR regulations with an analytical look at their effects on global operations in the chemical industry.

The Regulations

The REACH regulation covers requirements for chemical handling accountability by all sectors of the chemical industry supply chain within the European Union. These include manufacturers, exporters and importers of chemical substances, producers, exporters and importers of products or articles that contain chemical substances and end users of these substances and products. As the name indicates, the regulation mandates that decision making personnel evaluate chemicals, products and company operations for applicability to REACH and register said item(s) with the governing agency (ECHA). Registration includes providing classification of the chemical substance(s), detailed physical, chemical and toxicological data on the chemical substance(s), production/preparation method(s) and annual rates, description of the use(s) of these substances, a hazard assessment report (where mandated) and information on the company, site(s) and responsible contact person(s).

The IUR requirements, currently still a subpart of the TSCA Chemical Inventory regulations, stipulate that manufacturers, importers and users of chemical substances determine whether their operations are subject to the regulations and, if so, file periodic reports with the governing agency (USEPA). These reports include information on physical and chemical

Table 1 – Overview of REACH and TSCA IUR

	REACH	TSCA IUR
Purpose	• Collect generated data on regulated substances • Establish system to encourage / ensure viable substitution of substances posing high risk(s)	• Collect data on regulated chemicals and mixtures • Identify and manage risk to people and the environment
Administering Agency/Region	European Chemical Agency (ECHA) / European Union	United States Environmental Protection Agency (USEPA) / United States territories
Regulation Citation	Regulation (EC) 1907/2006	40 CFR 710 Subpart C (Proposed – 40 CFR 711)
Effective Date	December, 2006	December, 2006 (proposed *2011*)
Who Is Subject	Manufacturers, exporters and importers of regulated substances and producers/ importers of items containing these substances	Manufactures, importers, (proposed: and *users and processors*) of regulated substances for commercial use
Regulated Substances	Phase-In substances (including those listed on EINECS) except specific exemptions	Master Inventory File TSCA Section 8 (b), excluding exemptions
Threshold Quantities	≥1 Ton (metric); ≥10 tons; and ≥100 tons per calendar year	≥25,000 lbs (11,340 Kg); and ≥300,000 lbs per calendar year
Requirements Summary	Submit registration of substance • Administrative information - Name of registrant - Contact information - Location of site • Classification of substance per regulation • If in articles, weight percent • Physical / chemical properties and toxicity of substance • Documentation of testing • Brief description of use • Guidance for safe use •Tonnage range of annual production/import	Submit report for calendar year prior to reporting period • Administrative Information - Certification statement signed by authorized official - Company and Parent Company Name(s) - Site address • Chemical name / classification •Quantity manufactured and imported on site for principal reporting year (proposed: *and each year since last report*) • Details for principal reporting year (conc., worker exposure etc)
Format for Submittal	Electronic IUCLID format	e-IURweb reporting software

properties of reported chemical substances, production and usage at and importation to the site(s) for the principal reporting year, company and site addresses and responsible official contact details. The submitter is also required to report the estimated total number of workers likely to be exposed to each regulated substance at each site reported.

The proposed amendments to the IUR regulations require more detailed and specific information on the properties and categorizations of chemical substances reported, additional information on worker exposure and releases to the environment, details of the states, concentrations and quantities of these substances if/when they are sent off-site (i.e. exported to other US territories), specifics on processing and usage by the importing entities and information for each calendar year from the last reporting period (in addition to the principal reporting year). The frequency of reporting is also reduced from every five (5) to every four (4) years.

Regulatory Goals and Intents

The preamble to the REACH regulations gave a brief background on previous directives and laws governing chemicals that preceded the REACH regulation, which, to a great extent, incorporates and expands on those requirements. REACH encompasses chemical substances inventory requirements, chemical testing and categorization procedures, human health and safety assessments, environmental preservation measures and the maintenance of profitable commerce. The primary goal of the regulation is to focus the stakeholders of the chemical industry on all key aspects of their operations and to gather enough data so that well informed decisions can be made towards more viable, sustainable operations.

Similarly, the preamble to the proposed IUR amendments also gave background information on previously promulgated regulations governing chemicals, primarily TSCA and its Chemical Substance Inventory rule. While TSCA and the Inventory rule focus on assessing chemical substances and maintaining information on their manufacture/importation and movement through industry, and the current IUR regulations aim to collect and monitor updates to these data, the proposed regulations seek to effect a more comprehensive, detailed data collection process. The stated goal is to acquire the level of data that will foster more effective worker safety and environmental protection decision making

A Network of Regulatory Requirements

REACH and TSCA IUR are two of several regulations governing some aspect of the management of chemical substances and products by various stakeholders in the industry. As was stated earlier, currently effective IUR requirements constitute a subpart of TSCA which includes the requirements for initial chemical substances inventory and hazards/risk evaluation and management reporting. The USEPA administered Clean Air Act (CAA) also requires chemical hazards/risk evaluation and reporting in a Risk Management Plan (RMP), and similarly the Occupational, Safety and Health Administration's (OSHA's) rules contain a requirement for companies to perform chemical substances hazard analyses and include them in a Process Safety Management Plan.

These regulations clearly contain overlapping requirements. A comprehensive approach to complying with these regulations would result in the development of the basis for a chemical management system. Similarly, the fairly recent regulations governing the classification, labeling and packaging of chemicals (CLP rule) in Europe include requirements which overlap with some REACH requirements. Companies located in Europe and/or those doing business in Europe would find it expedient to use an integrated approach to comply with these two rules. Further, this approach could be incorporated into a broader management system that interconnects the requirements of all related regulations of REACH and TSCA IUR.

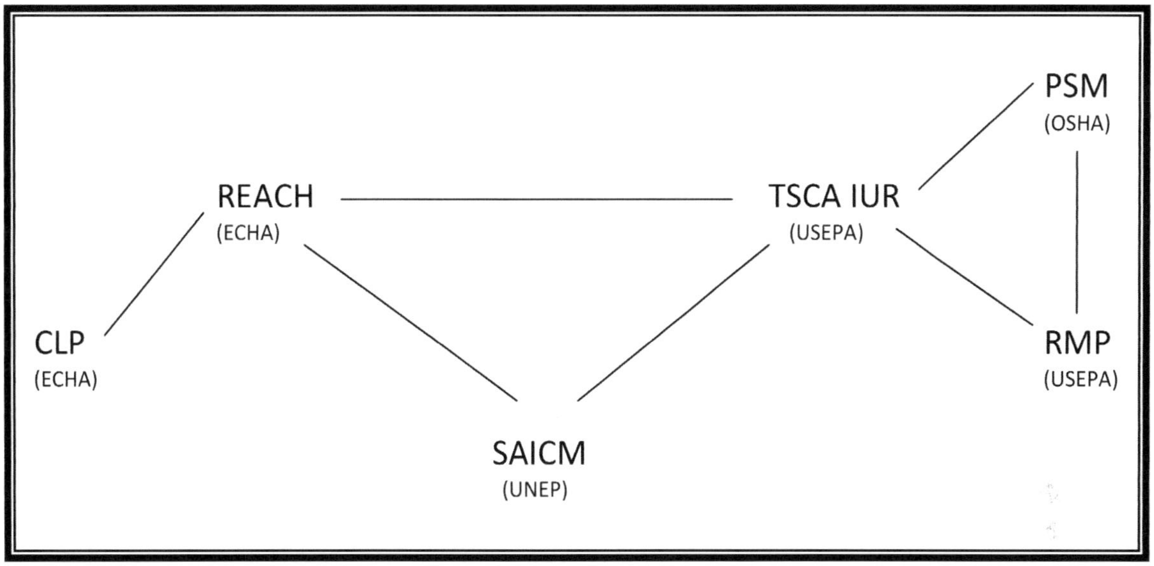

Fig.1. – Interconnection of Regional and Global Rules Incorporating Chemical Management Policies

In addition to national and regional rules, there are also international initiatives that govern the management of chemical substances. The United Nations Environmental Programme (UNEP) led International Conference on Chemical Management (ICCM) adopted a policy on the Strategic Approach to International Chemicals Management (SAICM). This provides a framework for developing and implementing sound lifecycle chemical management practices. Figure 1 gives a network representation of regional and global related rules and initiatives on the management of chemical.

Effects On Global Operations

Both REACH and TSCA IUR regulations contain similar core elements, like providing an overseeing agency with detailed information on the regulated substance(s) at pertinent sites, specifics on manufacture and processing operations, import and export data, usage of these substances and results of human health and safety evaluations (refer to Table 1.). These data are primarily collected and stored electronically, which makes them more easily assessable and transferable. This fact coupled with the level of detail required to comply with these regulations may raise a specter of concern for several members of the industry regarding increased risk of confidentiality breaches, despite the implementation of electronic technology safeguards. There is also the concern about resources required to generate, compile, document and submit the extensive information required, and maintain records for subsequent updates.

The dimension of business to business accountability may also be viewed as a burdensome responsibility. Both sets of regulations require that manufacturers and exporters of a regulated substance submit information on their products' uses and applications in various commercial arenas. Although the regulations offer coding guidelines, for example for usage and concentration ranges, it could be difficult to cite accurate descriptions since some chemical applications can be very unique to a chemical process. An exporter may justifiably have reservations affirming use and applicability by an importer.

If importer and exporter are in different regions of the world, they need to verify compliance requirements for applicable regulations. REACH was developed for the countries of the European Union and TSCA IUR regulations were developed for the United States of America and its territories, but each is capable of impacting commercial operations in the others region. A company with business operations in both regions will have personnel familiar with both sets of regulations and so will have the advantage of hands on working knowledge of both systems. A company with no direct business affiliation may need to expend significant resources in order to fulfill compliance of applicable requirements.

Internal Business Opportunities for Operational Enhancements

Despite the challenges of meeting applicable regulatory requirements, there are companies that have successfully devised and implemented strategic chemical management programs that have aided with compliance and enhanced operations. Dow Chemical Company (Dow) has been recognized for its implemented Chemical Management Policy which is geared towards achieving safe and environmentally sound chemical products manufacture, distribution, handling and disposal. The policy, built around five key principles of chemical manufacturing systems (outlined in Table 2. below) effectively addresses fundamental elements of REACH and TSCA IUR, as well as other related regulations.

Table 2. – Adaptation of Dow®-Promoted Chemical Management Principles

Principles of Chemical Management Systems

- **Risk-Based -** consider the intrinsic hazard(s) of a substance and the potential for these hazards to be effected; develop/implement related risk management strategy

- **Screen All Chemicals (new and existing) -** to determine further information needs in supporting a tiered risk-based approach to hazards and exposure management

- **Initially Leverage Available Information -** approach that encourages identification and evaluation of available information to decide the need for new testing

- **Recognize Shared Responsibilities Of Each Party Within Value Chain** – all stake holders need to assist in ensuring that chemicals are managed safely

- **Promote Transparency** – information is essential for making sound decisions about the safe management of chemicals

The company's chemical management initiative is one example of the actualization of one of many viable internal corporate opportunities that can be garnered from employing a comprehensive approach to chemical substances regulatory compliance. This holistic approach, while enabling the company to effectively achieve comprehensive compliance with all applicable regulations, creates a unique opportunity for assessment of internal operations which can reveal areas that could be improved for internal operational efficiency and overall company performance. Positive results in these areas contribute significantly to positive corporate image. Dow earned and award from the National Safety Council for its operational safety improvements.

Another chemical industry giant that has incorporated a comprehensive systems approach to chemical management is BASF. As part of a sustainability framework, the company has developed an eco-efficiency analysis (EEA) tool to comprehensively and comparatively assess optional products and processes. The EEA provides a methodology for assessing a product's/process' sustainability value using two of the three pillars of sustainable development, ecological/environmental preservation and economic gain. As show in Figure 2 below, an EEA's simultaneous consideration of environmental and economic factors yields a unique set of values per product or process which allows for direct efficiency comparisons.

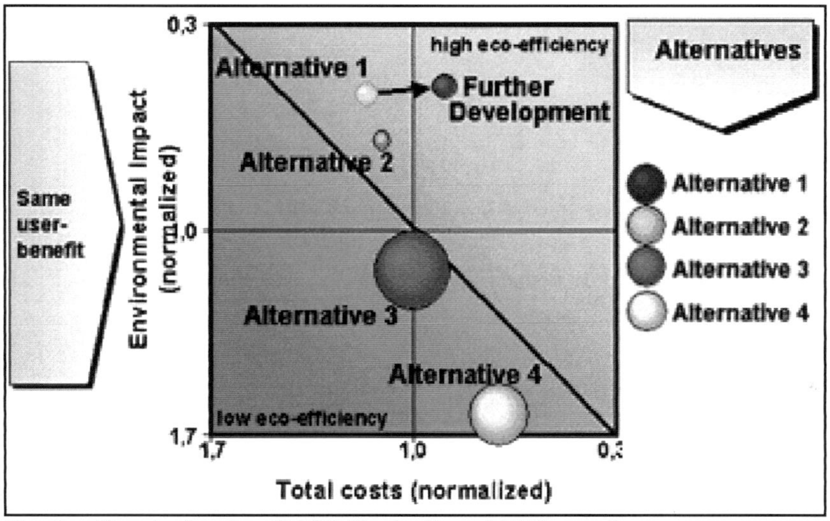

Fig. 2 Graph showing BASF Illustration of EEA results

The environmental assessment takes into account the lifecycle of the products and processes which includes the environmental impact of the materials used, the product uses and end of life disposal or recycling. It is clear that this assessment will invariable include compliance requirements of REACH and TSCA IUR, as well as other related regulations. Table 3 shows the six categories of environmental impact considered in BASF's EEA tool.

The two industry examples demonstrate the chemical industry's awareness of immediate and potential future benefits of applying chemical management system in business development and operation. They also reveal the groundwork for a new era in industrialization, where the early design stage for a product (or process) involves input from all stakeholder and product lifecycle and sustainability assessments.

Table 3. – Environmental Assessment - BASF EEA

Six Categories of Environmental Impact Assessment

- Raw Materials Consumption
- Energy Consumption
- Land Use
- Air and Water Emissions and Solid Waste Generation
- Potential Toxicity
- Potential Risk

External Business Opportunities

In addition to integrating and incorporating related regulatory requirements to create a comprehensive chemical management system and optimize resources, companies can take advantage of corporate or business opportunities To overcome difficulties in attaining regulatory compliance in a foreign territory, companies doing business across regions can explore the prospects of forming strategic alliances. A company located or based in a particular global region would be more knowledgeable of the letter and intent of the regulations of that region and would therefore be better equipped to advise on compliance options and procedures. This creates an opportunity to leverage that capability in a strategic relationship with a partner based in another region.

Furthermore, as noted previously, the proposed amendments to TSCA IUR show several similarities with REACH requirements, and several of these requirements are directed towards encouraging members of the chemical industry to develop and conduct business from a sustainability perspective – striking an optimized balance between profitable operation, human health/safety and sustaining/improving the environment. The SAICM initiative also includes an emphasis on safety as a component of sustainability on a global scale. In exploring and devising means of fulfilling these regulatory requirements, companies may also simultaneously take a critical look at various components of business operations for areas where change could, not only bring about ease of regulatory compliance, but also sustainability improvements which has been shown to augment business performance.

REFERENCE SOURCES

Official Journal of The European Union; 30.12.2006
www.regulations.gov
www.unep.org
www.dow.com
www.basf.com

Sustainability Indicators - a Review of the Literature

Helena Chum[(1)], Ethan Warner[(2)], Yimin Zhang[(2)], Garvin Heath[(2)], Maggie Mann[(2)], Alison Goss Eng[(3)], Goeran Berndes[(4)] and Kim Pingoud[(5)],
[(1)]National Renewable Energy Laboratory, Golden, CO, [(2)]Strategic Energy Analysis Center, National Renewable Energy Laboratory, Golden, CO, [(3)]Office of the Biomass Program, U.S. Department of Energy, Washington, DC, [(4)]Energy and Environment, Chalmers University of Technology Sweden, Göteborg, Sweden, [(5)]VTT, Helsinki, Finland

Abstract

Biomass for energy is an issue that crosses many sectors including climate, energy, land-use, and development policies (Marland et al. 2008). In any given context, limiting resources may define the extent to which land- and biomass- based energy can contribute to climate change mitigation (Schlamadinger et al., 2005). Biomass resources are geographically complex and multiple biomass resources may be used in a wide array of competing bio-based systems. Therefore, comparing the performance of different end-use bioenergy technologies to each other and relative to a reference system they displace are difficult and multivariate problems. Many impact indicators of bioenergy performance regarding environmental, economic and social considerations have been developed to help determine the optimal use of biomass resources. However, current policy approaches seeking to facilitate choices of technologies, products, and locations to reduce greenhouse gas emissions (GHG) have often focused on a limited number of indicators for bioenergy evaluation. Indicator selection has also been inconsistent across policy regimes. To compare the relative attractiveness of various bioenergy applications in reducing GHG emissions, multiple sustainability indicators should be employed to understand the advantages, disadvantages, and tradeoffs of different current and future bioenergy options within local/regional contexts.

Existing studies that evaluate the environmental impacts of bioenergy systems generally only report data to calculate one or two GHG emission indicators, with few studies that seek to deliberately evaluate multiple indicators. Currently, GHG emissions are often evaluated based on energy related indicators such as g CO_2 equivalent (g CO_2e) per MJ or per kWh in absolute terms and then compared to fossil fuel replacements to determine relative indicators. A common energy based relative indicator, ***relative GHG emission savings*** (% saved, with respect to the fossil alternative – transport fuel, heat, electricity, or combined heat and power), is typically calculated by comparing GHG emissions per unit of energy, or in the case of transportation the distance travelled, as it relates to the fossil fuel replacement. However, these two indicators do not capture effects on the resource (e.g., land, biomass) utilization efficiency of different bioenergy systems. Several such indicators emphasizing non-energy resources exist.

The ***displacement factor***, an absolute dimensionless indicator, deals with carbon resources and describes the reduction in GHG emissions from the displaced energy system per unit of biomass used (e.g., ton of CO_2e reduced per ton of C in biomass). Another resource related indicator, ***GHG savings per ha [m^2 or km^2] of land,*** describes GHG emission savings per unit of land used to produce biomass. This indicator can be used to determine which bioenergy options lead to optimal land use land to achieve GHG emissions reductions. Another indicator is the ***GHG savings per monetary unit input*** – either to the firm investing, thus tends to favour the lowest cost commercially available bioenergy options, or governments that seek to compare multiple options for investment and may add social and other considerations.

Several governments and non-government institutions have sought to reduce GHG emissions by replacing fossil energy sources or foster low GHG emissions bioenergy options by establishing policy systems based on GHG emission indicators. For example, the Renewable Energy Directive (RED) uses the relative GHG emission savings indicator (European Commission, 2009). The Environmental Protection Agency (EPA) of the US government adopted a similar approach that uses a relative indicator to set performance classes that meet the policy's biofuel production requirements (EPA, 2010). The California Air Resources Board (CARB) in the US is using an absolute indicator (i.e. g CO_2e/MJ) in their low carbon fuel standard which the transportation sector must meet (CARB, 2009). There are also several roundtable groups, such as the Roundtable on Sustainable Biofuels (RSB), that are progressing to establish accreditation systems based on indicator systems of various complexity, including social (Bauen et al. 2009; van Dam et al. 2010).

The application of a single indicator evaluating only one dimension (i.e., commonly energy and GHG emissions) across many contexts could have distorting effects on other environmental, economic, and social systems. Broader and longer term policy goals may be more difficult to achieve due to the lack of consideration of other factors such as the land availability or the biomass utilization efficiency. For example, results calculated from a Finnish study show how the displacement factor and GHG emission savings factor differ in the evaluation of various bioenergy systems (Soimakallio et al. 2009). The use of logging residues in modern combined heat and power (CHP) plants receives a relatively high ranking in GHG emission savings of about 86% (60-94% uncertainty range) regardless of the fossil replacement (e.g., coal, natural gas, etc.). Consideration of GHG savings from Fischer-Tropsch diesel produced in an integrated plant that minimizes electricity shows a nearly similar reduction of 78% (69-89% uncertainly range). In comparison the displacement factor is 78 t CO_2e/ t $C_{biomass}$ when coal is displaced and 30 t CO_2e/ t $C_{biomass}$ when natural gas is displaced using logging residues in a CHP plant. In contrast to the GHG emission savings indicator, Fischer-Tropsch diesel production from an integrated plant minimizing electricity is more likely to have a higher displacement factor (45 to 57 t CO_2e/ t $C_{biomass}$ uncertainty range).

The displacement factor provides investigates another dimension of bioenergy system's relation to carbon stocks that can be used to make a more informed decision. The use of a bioenergy route with a displacement factor less than one would just increase the net emissions (with respect to the carbon sink baseline) at least within the near term. Although the bioenergy route could be sustainable it cannot be considered climate or emission neutral and would require even more fossil C reductions to meet a set of emissions targets.

There are other issues with using only one indicator at a broad policy level. Using different single indicator approaches in multiple policies could negatively distort the relationships between various international systems (e.g., trade). Similarly, incorrect or mixed messages could be inferred from comparisons across policy regimes that lack parallel indicator systems or, in the case of relative indicators, parallel reference (i.e., fossil fuels) systems.

A comprehensive examination of the contextual issues/conditions that characterize large-scale biomass energy is an essential part of informed decision making. The method of applying multiple indicators is an important tool for policymakers because the geographic complexity of bioenergy resources with multiple uses implies many potential options and path-dependent routes that can be taken at multiple societal levels. Multiple indicators are necessary to understand the various scales of advantages, disadvantages, and tradeoffs of different current and future options to truly compare the ability of various bioenergy applications in reducing GHG emissions. Consideration of the local context in conjuncture with multiple indicators can make the analysis more complete and facilitate decision making at all levels.

Acknowledgement

Acknowledgments: The authors gratefully acknowledge André Faaij of Utrecht University, José Moreira of the Brazilian Biomass Reference Centre, the US Department of Energy Office of Energy Efficiency and Renewable Energy and the Intergovernmental Panel on Climate Change Working Group III, for discussions during the preparation of the Special Report on Renewable Sources of Energy and Climate Change Mitigation and Authors of the various chapters.

References

Bauen, A., F. Vuille, P. Watson, and K. Vad, 2009: The RSB GHG accounting scheme. Feasibility of a meta-methodology and way forward. . E4tech version 4.1, 91pp.

California Air Resources Board (CARB), 2010: Proposed regulation to implement the low carbon fuel standard. California Environmental Protection Agency AIR RESOURCES BOARD Stationary Source Division.

Environmental Protection Agency (EPA), 2010: Renewable Fuel Standard Program (RFS2) Regulatory Impact Analysis. EPA-420-R-10-006, Environmental Protection Agency.

European Commission (EC), 2009: Directive 2009/28/EC of the European Parliament and of the Council of 23 April 2009 on the promotion of the use of energy from renewable sources.

Marland, G., and M. Obersteiner, 2008: Large-scale biomass for energy, with considerations and cautions: an editorial comment. Climatic Change, 87(3), pp. 335-342.

Schlamadinger, B., B. Bosquet, C. Streck, I. Noble, M. Dutschke, and N. Bird, 2005: Can the EU emission trading scheme support CDM forestry. Climate Policy, 5, pp. 199-208.

Soimakallio, S., T. Mäkinen, T. Ekholm, K. Pahkala, H. Mikkola, and T. Paappanen, 2009b: Greenhouse gas balances of transportation biofuels, electricity and heat generation in Finland--Dealing with the uncertainties. Energy Policy, 37(1), pp. 80-90.

van Dam, J., M. Junginger, and A.P.C. Faaij, 2010: From the global efforts on certification of bioenergy towards an integrated approach based on sustainable land use planning. Renewable and Sustainable Energy Reviews 14(9), pp. 2445-2472.

(99b) Sustainability Metrics for Biofuel Production Via Thermochemical and Biochemical Conversion Platforms

Eric C. D. Tan and Daniel Inman, National Bioenergy Center, National Renewable Energy Laboratory, Golden, CO

The National Renewable Energy Laboratory (NREL) is developing technologies to convert biomass to biofuels via biochemical and thermochemical conversion processes. Three environmental sustainability metricsógreenhouse gas (GHG) emissions, consumptive water use, and fossil energy demandóare quantified using process engineering models. Currently, NREL is working to identify technology improvements in the conversion processes that will reduce these environmental emissions. This effort will improve our understanding of the environmental tradeoffs between conversion technology options. For example, in the NREL biochemical conversion process design, ammonium hydroxide has replaced overliming as for conditioning the hydrolysate. The GHG tradeoffs associated with this technology change will be examined. Aspen Plus software is used to evaluate process enhancements and to quantify the emissions, as well as to calculate the cost of such a reduction. Furthermore, developing annual state-of-technology (SOT) sustainability assessments for the conversion platforms is critical to the overarching goal of developing a sustainable biofuels industry. Both the current SOT and process modifications to the current technology have been assessed and will be updated based on the latest conversion SOT for their relative sustainability.

See more of this Session: Establishing and Ensuring Sustainable Bioenergy Development
See more of this Group/Topical: Topical 3: Advances in Sustainability

(99c) Sustainable Supply Chain Design: Biofuel Impacts, Indicators, and Metrics

Raymond L. Smith, US EPA, Cincinnati, OH

A project within the U.S. EPA's Office of Research and Development is studying biofuel and petroleum supply chains. The project is first considering corn-to-ethanol as the biofuel system of interest, with other systems studied in the future. The analysis of the system will consider many environmental impacts which are included in the EPA's Tool for the Reduction and Assessment of Chemical and other Impacts (TRACI). Other analysis results will provide indicators and sustainability metrics to describe the system. These sustainability metrics will include cumulative exergy and emergy, ecological footprint, renewability, and water use. Additional metrics are being considered. A second part of the work will consider the design of supply chains for improved sustainability. The synthesis of these systems, along with analysis, will suggest potential improvements to future biofuel supply chains. This presentation will provide a progress report and suggest directions of further work.

See more of this Session: Establishing and Ensuring Sustainable Bioenergy Development
See more of this Group/Topical: Topical 3: Advances in Sustainability

The RSB Certification System and the RSB's Multi-Methodology Tool for Assessing Biofuel Sustainability

Junquera, V. [1], Guiramand, M. [1], Kopse, A. [1], Faist Emmenegger, M[2], Reinhard, J. [2], Zah, R. [2], Ziep, T. [3], Widok, A. [3], Wohlgemuth, V. [3]

[1] Roundtable on Sustainable Biofuels, CH-1015 Lausanne
[2]Empa, Technology and Society Lab, Life Cycle Assessment and Modelling Group, CH-8600 Duebendorf
[3]HTW Berlin, DE-12459 Berlin

The Roundtable on Sustainable Biofuels (RSB, http://rsb.epfl.ch) is an international standard for sustainable biofuel production developed through an open & multi-stakeholder process involving social and environmental NGOs, government agencies, inter-governmental organizations, and private companies involved in biomass and biofuel production. During the nearly four years that it took to develop the RSB Standard, the RSB held numerous regional consultations across the globe with over 900 participants from more than 15 countries; the RSB Standard was developed through a consensus-based process by the RSB multi-stakeholder membership. In November 2010, the RSB has released "Version Two" of its standard. RSB plans to start certifying biofuel operations in the first quarter of 2011.

As part of its efforts to develop a user-friendly and practical certification system, the RSB has developed an online compliance tool, the "RSB Tool", in conjunction with the Swiss Federal Laboratories for Materials Science and Technology (EMPA, www.empa.ch) and with the technical assistance from the University of Applied Sciences Berlin (HTW Berlin, http://www-en.htw-berlin.de). The RSB Tool is originally based on the Sustainability Quick Check for Biofuels (www.sqcb.org), which is a web-based, flexible tool to assess the sustainability of biofuels.

The RSB Tool is an online compliance-check and GHG calculation tool of the RSB. It will be freely available online once released in the first half of 2011. Target users are participating operators of the RSB certification systems or other interested persons or organizations. The tool serves several purposes, including information, evaluation, and guidance.

The tool serves as an entry door to the RSB certification system by providing information on the RSB certification process and allowing the user to conduct self-evaluations against the RSB Standard. The user can check compliance with the RSB Principles & Criteria and Indicators and perform a Self Risk Assessment according to the RSB Standard for Risk Management.

In addition, the tool has built-in calculations or values of lifecycle GHG emissions of biofuels following various international methodologies, including the RSB GHG lifecycle methodology, the European Union Renewable Energy Directive (EU RED) methodology, the Swiss ordinance on mineral oil tax, the California Low Carbon Fuel Standard and the U.S. Renewable Fuel Standard (RFS2/USA). The RSB GHG methodology is a lifecycle GHG calculation methodology for Biofuels developed in conjunction with EMPA and available on the RSB website.

For the RSB and EU RED methodologies, the tool uses a modular approach that allows the user to calculate GHG emissions for their scope of operations within the biofuel chain of production. Moreover, the user enters actual values reflective of their operations, including data on material and energy usage. Available default data allow the operator to assess missing parts of the production chain and compare his performance to all globally relevant benchmarks.

(99e) Bioenergy Knowledge Discovery Framework: An Integrated Platform Supporting Nation's Bioenergy Infrastructure

Budhendra Bhaduri, Aaron Myers, Amy Rose, Philip Nugent and Rahul Ramachandran, Oak Ridge National Laboratory, Oak Ridge, TN

Introduction

Biomass-derived fuels have the potential to be one possible solution to our nation's dependence on imported oil. In 2007, the Energy Independence and Security Act (EISA) was passed with the aim to increase the supply of alternative fuels by requiring transportation fuel sold in the U.S. to contain a minimum of 36 billion gallons of renewable fuels by 2022. This included advanced and cellulosic biofuels and biomass-based diesel,. In response, the U.S. Department of Energy (DOE) recognized the need for having a diverse energy portfolio to meet the nation's energy security challenges and has set up a strategic plan to promote energy security through a varied energy supply that is reliable, clean, and affordable. The DOE Office of Biomass Program supports these objectives by trying to establish a viable domestic bioindustry. However, establishing a domestic bioindustry depends on several factors. First, significant and rapid advances in biomass feedstock and conversion technologies must take place. Second, there should be large volumes of sustainable biomass feedstock to sustain the industry. Third, large- scale, integrated biofuels production facilities need to be deployed. And the finally, there should be an adequate biofuels infrastructure to support this industry.

Since the biofuel infrastructure is complex with several critical components such as energy, transportation, water, agricultural and commerce, the need for an integrated decision support tool and collaborative platform was felt in order to support the different stakeholders. The Bioenergy Knowledge Discovery Framework (KDF) is currently being established at the Oak Ridge National Laboratory (ORNL) in response to this need. The Bioenergy KDF seeks to harmonize data from different sources and provide an integrated decision-making capabilities to its stakeholders.

Architecture Overview

A Geospatial KDF tool being developed at ORNL powers the Bioenergy KDF. The GeoKDF tool is a general-purpose tool focused on cataloging, sharing, visualization, mapping, modeling and analysis of geospatial data. The GeoKDF tool can be instantiated for specific domains such as the bioenergy. The GeoKDF has flexible client-server architecture. The server side consists of spatial database and a data storage server. Both spatial and non-spatial data can be stored on the server. The client queries the stored data and the spatial datasets are served as maps, features or coverages using standards-based specifications via a Geoserver. The GeoKDF provides the client functionality via the web browser to improve accessibility. The browser-based client is written in Javascript and utilizes both the JQuery and the ExtJS library to provide rich interactive functionality to the end user. The visualization of maps and layers within the client is achieved by using OpenLayers API.

Stakeholders and Use case Scenarios

The Bioenergy KDF is being configured to meet the functional requirements of its four stakeholders.

First, the framework provides *policymakers* at both federal and state level access to both the latest research data on biomass production and a decision support tool to make strategic decisions. The policymaker could use this framework to ascertain the amount of feedstock currently available within a specific state. Does a region have any crop residue feedstock? How much forest residue feedstock is available for this region? Policymakers can pose such queries to the framework. In addition, the policymakers can utilize the Bioenergy KDF to look at future projections of energy crop potential for specific timeframes. These querying capabilities will allow policymakers to make informed decision towards aligning strategic policies toward renewable energy solutions.

Second, the framework provides the *private industry* both the information and actionable knowledge to make business decisions and exploit business opportunities. Latest research data on biomass feedstock production, future projections and transportation models are accessible via the Bioenergy KDF. Business ventures could use the framework to explore the viability of building a Biorefinery near a certain city and look at the availability of feedstock nearby to minimize transportation costs. The Bioenergy KDF could be used to find suitable counties with feedstock production greater than certain threshold to meet the requirement of a proposed Biorefinery of a certain capacity.

Third, the Bioenergy KDF supports *scientists and researchers* by being much more than a simple web based decision support tool. In fact, it is designed to be a comprehensive online collaborative tool for information sharing. Scientists and researchers from national laboratories, universities and private industries can register and get accounts on the framework. Once registered, the scientist and researchers are able to upload files, edit information online and exchange knowledge with others within and across domains. The framework provides the scientists full control of their data and information, allowing them to decide with whom to share their expertise. In addition, the framework serves as a platform for these scientists and researchers to publicize, results and expertise and connect with the consumers of such information.

Fourth, the Bioenergy KDF, being openly available on the web, serves an important outreach objective of informing the *general public* about the ongoing efforts towards building nation's renewable energy program. For example, high school students can use the framework to explore the data by creating simple maps. They can explore the role their county or state will play in providing feedstock to Biorefineries or locate all the bioenergy facilities in their state. They could find locations of other biomass sources such as pulp/paper mills, coal fired plants near them and explore future projections for energy crop for their area.

Summary

The Bioenergy KDF is currently being developed to provide an integrated decision support tool and collaborative platform in order to support nation's Biofuel infrastructure. The Bioenergy KDF will provide functionality to share, visualize, map, model and analyze geospatial and non-geospatial data from multiple repositories. The functionality of the Bioenergy KDF is being designed to address the unique needs of different stakeholders. These stakeholders will play a critical role in achieving our nation's biofuel based renewable energy objectives.

References

Biomass Multi Year Program Plan, Office of the Biomass Program, Energy Efficiency and Renewable Energy, U.S. Department of Energy, August 2007

See more of this Session: Establishing and Ensuring Sustainable Bioenergy Development
See more of this Group/Topical: Topical 3: Advances in Sustainability

(99f) Life Cycle Assessment Tools for Algal Biofuels

Edward Frank[1], Amgad Elgowainy[2], Ignasi Palou-Rivera[3], Michael Wang[4] and Jeongwoo Han[3], (1) Energy Systems, Process Technology Research Section, Argonne National Lab, Argonne, IL, (2) Argonne National Lab, Argonne, IL, (3)Energy Systems Division, Argonne National Laboratory, Argonne, IL, (4)Energy Systems, Argonne National Laboratory, Argonne, IL

Life Cycle Assessment Tools for Algal Biofuels

Algal biofuels are being pursued aggressively to meet Renewable Fuel Standards mandates for advanced biofuels. A life cycle assessment (LCA) of emissions is required to qualify as an advanced biofuel. Several LCA studies have been published for a few particular approaches; however, it is difficult to compare between technologies because of differences amongst inputs, system boundaries, and analysis approaches in these seminal studies. The Greenhouse Gases, Regulated Emissions, and Energy Use in Transportation Model (GREET) developed at Argonne National Laboratory has analyzed many biofuel and petro- fuel pathways. We report here work to enable systematic comparisons of algal biofuel pathway options. The work comprises two elements 1) An algal process model supporting a growing number of unit operations that is able to encode processing from CO2 transport through conversion to fuel and 2) extensions to GREET that enable detailed life cycle analysis of algal biofuels and co-products. The project objective is to work collaboratively to provide life-cycle assessment tools to the broad community of academic, government, and industrial labs that can form a basis for uniform analysis of algae technology options.

See more of this Session: Establishing and Ensuring Sustainable Bioenergy Development
See more of this Group/Topical: Topical 3: Advances in Sustainability

Anaerobic Treatment of High-Strength Wastewater in the Food & Beverage and Agricultural Industries

Carlos D. Claros, P.E., MWH Global, Inc., Chicago, Illinois
(carlos.d.claros@us.mwhglobal.com)

Introduction

Most industrial facilities in the agricultural and food & beverage sectors generate high-strength organic wastewater discharges that can be biologically treated by either aerobic or anaerobic means. Table 1 shows average raw wastewater characteristics of actual operating facilities. Within these sectors, some facilities, such as breweries and sugar processing plants, generate wastewater with high concentrations of organic matter and relatively low suspended solids (TSS <1,000 mg/L). Other facilities, such as corn wet mills, olive mills and rendering plants, present high concentrations of organic matter and solids. Wastewater from dairy, rendering and olive mill facilities typically contains a high concentration of fats, oils and grease (FOG). Although not shown in Table 1, nitrogen and phosphorous are other constituents of importance, typically present in meat processing and rendering plant effluents. Wastewater generated in all these industries has wide variability in both concentration and flow rate.

Table 1 – Raw Wastewater Quality from Representative Facilities

Sector	Facility	Flow (mgd)	BOD_5 (mg/L)	COD (mg/L)	TSS (mg/L)	FOG (mg/L)
Food &Beverage	Brewery	0.13	5,000	9000	600	20-50
	Dairy	0.03	1,700	2,500	1,020	500
	Salad Dressing	0.13	18,000	34,000	10,900	850
Agricultural	Olive Mill	0.06	11,000	43,000	8,000	1000
	Rendering	0.13	--	18,000	5,700	2,200
	Sugar Mill	1.6	--	7,600	< 500	--
	Corn Wet Mill	2.8	3,700	6,200	1,400	< 50

BOD_5: Biological Oxygen Demand (5-day); COD: Chemical Oxygen Demand

Biological treatment of high-strength wastewater has been traditionally addressed by either aerobic treatment, or lagoon systems including anaerobic lagoons, or combinations of both. Aerobic processes include activated sludge, trickling filters, moving bed bioreactors and aerobic membrane reactors. Lagoon systems consist of combinations of anaerobic, facultative and settling lagoons, typically in a series arrangement.

Anaerobic treatment of high-rate design presents the advantages of a smaller footprint and lower installed power than aerobic processes, resulting in lower capital costs when building a new treatment plant or retrofitting an existing plant. Also, no aeration is required and consumption of chemicals for nutrient supplementation is lower than that for aerobic processes, resulting in lower Operation and Maintenance (O&M) costs (Speece, 2008) (Driessen & Vereijken, 2003). An additional benefit is the production of biogas, a renewable energy attribute of this technology.

Despite these advantages, anaerobic treatment has historically been less attractive because it does not provide the complete level of treatment that can be achieved by aerobic means. In addition, the historical low cost of energy in the U.S. has overshadowed the benefits of biogas production and lower energy consumption of anaerobic processes.

This paper provides an overview of high-rate anaerobic technologies, discusses advantages of anaerobic high-rate treatment over aerobic treatment, and presents a case study in which cost-related parameters were compared vis-à-vis for treatment schemes with and without anaerobic reactors.

Treatment Requirements

Broadly speaking, wastewater treatment requirements are driven by whether the plant effluents are discharged to a Publicly-Owned Treatment Works (POTW), or a surface water body. If the facility discharges to a POTW, the POTW assesses effluent disposal costs generally based on flow rate, organic and solids loading. Under this scheme, treatment is centralized by the POTW that receives raw industrial wastewater from facilities connected to the sewer system. Nevertheless, if solids or FOG discharges exceeds the POTW capacity, the POTW may require the facility to provide effluent pretreatment prior to discharge.

On the other hand, if the facility discharges directly to a surface water body, a comprehensive wastewater treatment scheme is generally required, consisting of not only FOG and solids pretreatment, but also secondary treatment and post-treatment for nutrient removal. Effluent discharges to surface water bodies and waterways are regulated by stringent "National Pollutant Discharge Elimination System" (NPDES) regulations.

High-Rate Anaerobic Treatment Overview

Anaerobic treatment is a biological process that converts organic matter to volatile fatty acids, and subsequently to biogas. Biogas is primarily composed of 60 to 70% methane, and 25% carbon dioxide. Biogas can be used to fuel boilers and co-generation equipment such as engine generators, turbines, and fuel cells.

Anaerobic treatment has been widely used in digestion of municipal sludge and agricultural waste since early 1900s. In the U.S., its applicability in industrial wastewater treatment dates back to 1955, with the appearance of the anaerobic "Contact Process" (CP) developed by G.J. Schroepfer of the University of Minnesota.

One of the first successful full-scale applications of a "high-rate" reactor was reported in South Africa in 1961. This reactor was known as "clarigester" and was installed at a municipal waterworks to treat maize processing wastewater (Ross, 1989). The clarigester concept was later improved by G. Lettinga of Wageningen University, who developed the "Up-flow Anaerobic Sludge Blanket" (UASB) reactor in 1979, which has been widely implemented at numerous facilities with soluble organic wastewater (Lettinga & Hulshoff, 1991). High-rate anaerobic treatment has rapidly evolved in the last fifteen (15) years, and new advancements in anaerobic reactor technology are expanding their applications into other areas beyond end-of-pipe treatment (Driessen W. , 2001).

Table 2 summarizes the characteristics of anaerobic reactor types including the operating description, organic loading rates, and influent quality criteria. Anaerobic lagoons and contact process are categorized as low-rate and medium-rate reactors, respectively, with organic loading rates of between 0.1 and 5 kgCOD/(m^3.d), as shown in Table 2. Lagoons are widely used in agricultural, and meat and poultry facilities located in rural areas where land space is ample (Totzke, 2009). Contact process is suitable for slurry-type wastewater with very high COD (>70 g/L), solids, and FOG. Contact process has been successfully used for treating distillery wastes, and meat packing wastewater (Eckenfelder, 2000).

Table 2 - Anaerobic Reactor Types

Reactor Type	Acronym	Organic loading rate kg COD/(m^3.d)	Characteristics	Influent Quality
Anaerobic Lagoon	AnL	0.1 - 1.0	Flocculent biomass; 6 < Depth < 18 feet	High TSS, FOG,
Contact Process (pseudo-CSTR)	CP	1 – 5	Flocculent biomass, Mechanical agitation; External settler with Sludge recirculation	Slurry/sludge-type; Very high COD and TSS
Anaerobic Filter (AF)	AF	5 – 10	Attached and flocculent biomass, fixed-media bed, upflow/downflow	Soluble COD, Low to Moderate TSS FOG < 100 mg/L
Anaerobic Floatation Reactor	*AFR*	*5 – 10*	*Contact process with inner flotation unit; Flocculent sludge*	*Total COD High TSS and FOG*
Up-flow Anaerobic Sludge Blanket	UASB	5 – 15	Suspended granular biomass, Gas/liquid/solid separator; Hydraulic/biogas agitation	Soluble COD Low TSS (<10%COD) FOG < 100 mg/L
Anaerobic Membrane Bioreactor	*An-MBR*	*8 – 15*	*Combination of CSTR + Ultrafiltration membrane Flocculent biomass*	*Total COD High TSS and FOG*
Expanded Granular Sludge Blanket	EGSB	15 – 25	Expanded granular biomass; G/L/S separator High recirculation	Soluble COD Low TSS FOG < 50 mg/L
Internal Circulation	IC	20 – 30	Expanded granular biomass; Double G/L/S separator High recirculation	Soluble COD Low TSS FOG < 50 mg/L

The other anaerobic reactor types shown in Table 2 are categorized as high-rate reactors. High rate anaerobic reactors operate at organic loadings of above 5 kg COD/(m^3.d). A treatment scheme involving an anaerobic reactor commonly includes equalization to attenuate variability of influent quality, and to provide acidification of complex organic substrates (Lettinga & Hulshoff, 1991). If high concentrations of solids and FOG are anticipated, pretreatment may be required depending on the reactor type selected.

UASB was widely implemented from 1980 through mid-90s, accounting for over 50 percent of the high-rate system installation constructed in that period worldwide. The other 50 percent corresponded to expanded, fluidized, and fixed-film (AF) anaerobic reactors. From the mid-90s on, with more successful applications of expanded bed technology being reported, EGSB and IC reactors gained a larger share of the market. EGSB and IC are compact reactor units, suitable when available space is tight. Nowadays, over 60 percent of new high-rate projects are expanded bed technologies (Van Lier, 2008).

More recently, two new technologies have become available in the marketplace: An-MBR and AFR. Both technologies treat wastewater with high solids and FOG content. They are generally compact and the need for equalization is generally reduced. An-MBR consists of an anaerobic reactor followed by an ultra-filtration membrane. An-MBR has been used successfully in dairy, salad dressing and distillery waste applications. (Christian, 2009).

AFR is a novel anaerobic reactor including a central, inner flotation unit that separates FOG by an effluent recycle stream saturated with biogas. FOG is returned to the bottom zone of reactor for degradation. One full-scale plant is currently operating at an ice-cream factory in Europe, designed by

Paques. A similar reactor was reported in Uruguay for dairy wastewater, but this one requires a separate digester for treating floated material (Passeggi, 2009).

Case Study: Rendering Industry - Comparison of Aerobic versus Anaerobic

Waste minimization and treatment options were assessed at a rendering facility processing poultry and cattle by-products, including feathers, blood, grease and bones. The existing wastewater treatment plant consisted of a flotation unit and a system of stabilization lagoons prior to final disposal to surface water. Aluminum sulfate and polymer were continuously dosed at the inlet of the flotation unit. The existing treatment did not consistently meet the discharge requirements and was a source of odor nuisance. The raw wastewater flow averaged 88 gpm, with average concentrations of COD at 18,000 mg/L, TSS at 5,700 mg/L, and FOG at 2,200 mg/L.

Based on characterization and treatability testing, waste minimization actions were recommended at three manufacturing areas to reduce pollutant loading in the plant effluent. These actions consisted of improving centrifugation by pH adjustment and incorporating settlers to recover solids. Utilizing data from characterization and treatability, mass balance estimates were developed to assess the magnitude of the pollutant loading reduction. As a result, it was concluded that minimization actions would result in reductions of COD and TSS loading of approximately 50 and 80 percent, respectively.

Subsequently, a treatment alternatives assessment was conducted to evaluate treatment schemes incorporating high-rate anaerobic treatment versus aerobic treatment alone. The influent criteria utilized for this comparison were developed assuming that the recommended minimization actions would be implemented and an improved flotation process would provide sufficient solids and FOG pretreatment. Table 3 shows the resulting influent criteria.

Table 3 - Wastewater Quality after Minimization and Pretreatment

Parameter	Units	Concentration
Flow Rate	[MGD]	0.126
COD	[mg/L]	7,200
TSS	[mg/L]	290
FOG	[mg/L]	140
Temperature	°F	80 – 90

The following potential treatment alternatives were assessed:

Alternative 1 - Anaerobic reactor + lagoon in series
Alternative 2 - Activated sludge (extended aeration)
Alternative 3 - Anaerobic reactor + Activated sludge

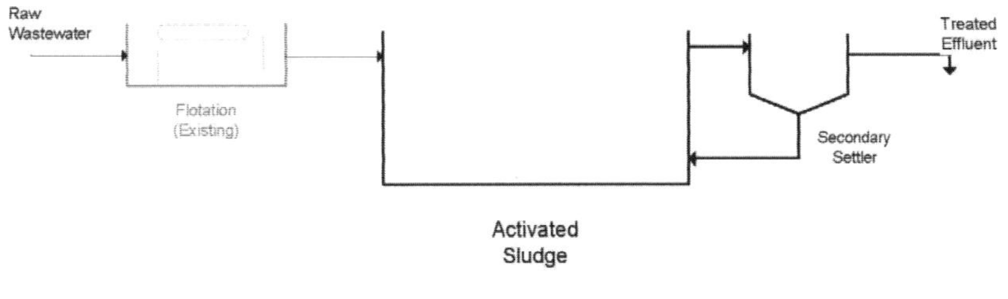

Figure 1 - Treatment Alternative "2": Activated Sludge (Extended Aeration)

Figure 1 shows the treatment scheme of Alternative 2. The activated sludge selected for the comparison corresponds to an extended aeration design. The secondary settler retains biomass that is recycled back to the reactor.

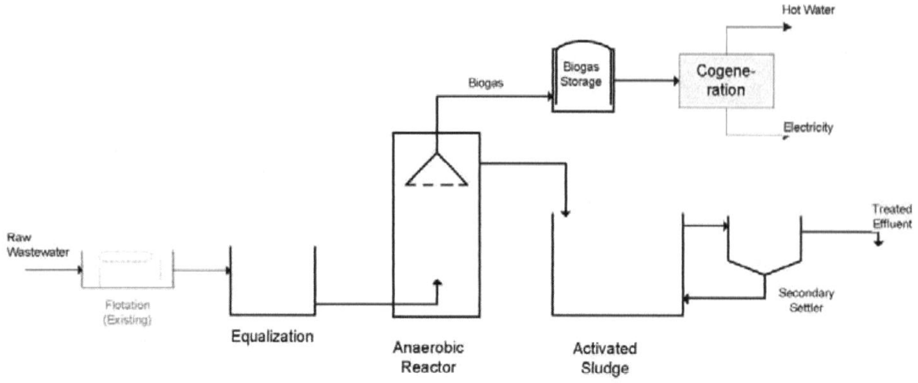

Figure 2 - Treatment Alternative "3" – Combined Anaerobic and Aerobic Reactors

Figure 2 shows the treatment scheme of Alternative 3 that consists of equalization, an anaerobic reactor, and an activated sludge process. Figure 2 also shows biogas storage and a cogeneration unit for beneficial use of biogas.

These treatment alternatives were evaluated in terms of capital and O&M cost-related parameters including energy use, size and footprint, and residual sludge. Table 4 shows the design and performance criteria selected for estimating these parameters.

Table 4 –Design and Performance Criteria Selected for Anaerobic and Aerobic Reactors

Criteria	Units	Activate Sludge[1]	Anaerobic Reactor
COD removal	%	98%	75 %
Equalization Residence Time	Hours	---	12
Organic Volumetric Loading	kg COD/(m^3. d)	1.5	10
Secondary Settler	Gal/(ft^2.d)	350	Inner to reactor
Energy – Aeration	kWh/ lb COD_{rem}	0.32	N/A
Sludge production	lb TSS/lb COD_{rem}	(0.1 – 0.25)	0.02 (0.01 – 0.03)
Methane production[2]	$scfCH_4/lbCOD_{rem}$	N/A	4.5
Methane heating value	mmBtu/scf CH_4	N/A	910
Temperature	°F	60 - 80	85 – 95
Nutrients (COD:N:P)	Weight basis	100:5:1	400:7:1

1- Same design criteria of activated sludge is used for both Alternatives 2 and 3
2- Methane production = theoretical production x 0.8
N/A: Not applicable;

Table 5 shows a subset of the data utilized for the comparison, specifically major capital and O&M parameters for Alternatives 2 and 3. One challenge identified during the study was that

Alternative 1 may not achieve complete nitrification, thus the comparative data presented in Table 5 does not include Alternative 1, and focuses on the more feasible alternatives.

Table 5- Comparison of Cost-related Parameters for Aerobic versus Anaerobic Treatment

Parameter of Comparison	Units	Aerobic Alternative 2	Anaerobic Alternative 3
Reactor Type		Activated Sludge (extended aeration)	Anaerobic Reactor + Activate Sludge
Capital Cost Items			
Equalization	Gallons	---	60,000
Reactor Volume	Gallons	610,000	Anaerobic: 92,000 Aerobic: 140,000
Secondary Settling Area	sq. ft.	360	360
Aeration Power Installed	HP	130	30
O&M Cost Items			
Aeration Power Consumption	kWh/day	2,400	550
Sludge Production	lbs TS (dry)/day	2,100	480
Biogas Production	ft^3/day	N/A	64,000
Energy Production	mmBtu/day	N/A	(+) 58
Reactor Heating	mmBtu/day	N/A	(-) 5
Alkalinity	lbs/day (as $CaCO_3$)	--	145

(-) or (+) represents energy consumption or production, respectively.

Energy Consumption

Alternatives 1 and 3, involving anaerobic reactors, would generate biogas with an energy content amounting to 58 mmBtu per day, as shown in Table 5. The anaerobic reactors operate in a temperature range of between 85 and 95 °F. Since raw wastewater temperatures at this rendering facility ranged from 80 to 90 °F, heating requirements were estimated at 5 mmBtu per day. As a result, biogas with a net energy value of 53 mmBtu per day would be available for beneficial use at the plant.

Since anaerobic treatment is generally not able to achieve surface water discharge requirements, post-treatment was included in Alternatives 1 and 3. In particular, activated sludge was included as post-treatment for Alternative 3, which would incur approximately 550 kWh per day of electricity usage.

Conversely, the sole use of extended aeration activated sludge in Alternative 2 would incur an electricity consumption of 2,400 kWh per day, or four times the energy required in Alternative 3.

Process Unit Sizes

As shown in Table 5, a comparison of total reactor volumes shows the combined anaerobic and aerobic scheme (Alternative 3) would require an aggregate reactor volume of 232,000 gallons, whereas the sole aerobic reactor option would require a much larger volume of 610,000 gallons (Alternative 2).

Since anaerobic system performance is sensitive to organic loading variability, an equalization volume equivalent to 60,000 gallons is needed for Alternatives 1 and 3. This would bring the total process volume required in Alternative 3 to 292,000 gallons, or about half the total process volume required for the activated sludge option.

Secondary settling associated with the activated sludge units are deemed to be of similar size for both alternatives, in turn they do not impact the comparative results in relative terms.

Sludge generation and disposal

A comparison of the daily sludge production demonstrates that the schemes incorporating an anaerobic reactor would generate a much lower amount of sludge. As shown in Table 5, Alternatives 1 and 3 would generate 390 and 480 pounds per day of sludge, respectively, while Alternative 2 would generate 2,100 pounds per day.

The sludge generated in the activated sludge reactor requires digestion and dewatering prior to land disposal, thereby incurring significant capital and O&M costs for sludge management. Anaerobic processes produce a much lower amount of sludge, thereby incurring much lower costs of sludge management.

Power installed

Installed power for aeration equipment is the major energy intensive component of aerobic systems. In contrast, anaerobic reactors do not require aeration, thus contributing significant capital cost savings. For instance, in this comparative analysis, Alternative 2 involving an extended aeration process requires a minimum installed power of 130 HP, while Alternative 3 including an anaerobic reactor in the treatment scheme requires 30 HP.

Common equipment used in both systems includes feed and recirculation pumps whose installed powers are assumed comparable, and do not contribute significantly to the relative comparison.

Chemical Consumption

Biological reactors need to meet certain nutritional and alkalinity requirements for bacterial growth and stability. Anaerobic reactors treating high-strength wastewater can operate at COD:N:P ratios of above 400:7:1 (weight basis); while similar aerobic systems require in average 100:5:1 as shown in Table 4. Since rendering effluents present relatively high concentrations of nitrogen (TKN) of between 500 to 1,000 mg/L (Sindt, 2006), capital and O&M costs for nutrient addition would be low for the alternatives presented but even more so for the anaerobic systems due to lower nitrogen and phosphorous requirements per pound of COD.

Regarding alkalinity, anaerobic reactors require alkalinity for pH control due to the acidic effects of carbon dioxide in the reactor biomass. Nevertheless, anaerobic metabolism can produce alkalinity from protein-rich substrates such as rendering effluents, in turn reducing significantly the need for alkalinity dosing. On the other hand, aerobic reactors treating effluents with high concentrations of nitrogen require alkalinity for nitrification.

In summary, chemical consumption for the alternatives presented is assumed to be comparable, and do not contribute significantly to this relative comparison.

Conclusions

The case study demonstrates that the incorporation of an anaerobic reactor in the treatment scheme resulted in the most attractive alternative due to advantages in energy costs, sludge generation, sizing, and installed power. As an additional benefit, biogas with a net energy value of 53 mmBtu per day would be produced, sufficient to power a cogeneration unit for producing hot water and electricity.

Food and beverage, and agricultural industries can benefit from implementing anaerobic wastewater treatment at their facilities by replacing or retrofitting existing treatment systems with high-rate anaerobic reactors. To this end, its financial feasibility must recognize not only capital and O&M cost savings, but also the full revenue of biogas utilization, and potentially of greenhouse emissions reduction.

Anaerobic high-rate technologies are widely used in high-strength wastewater primarily in the food and beverage industry, and to a lesser extent in the agricultural field where lagoons have historically seen more widespread use.

Wastewater characterization and treatability evaluation are important components of feasibility assessment of anaerobic treatment. Characterization efforts assist with determining opportunities for waste minimization at the process-area level by adjusting process operations, or by at-source pretreatment. Treatability testing assists with assessing wastewater anaerobic degradability, nutrient requirements, and toxicity effects, as well as, evaluating minimization opportunities at the plant level.

High-rate anaerobic treatment continues to evolve to address even more complex wastewater, and to reduce the need for pretreatment of effluents containing high solids and FOG. Pilot testing is required when no sufficient full-scale data is available for wastewater at hand, as is the case of the novel AFR and An-MBR.

In light of the increasing cost of effluent disposal in terms of discharge and surcharge fees, high-rate anaerobic treatment may serve as a cost-effective pretreatment means for many facilities discharging high-strength wastewater to sewers served by POTWs.

References

Christian, S. (2009). Anaerobic Membrane (AnMBR) Technology. *Anaerobic Treatment of High-Strength Industrial Waste.* Milwaukee: Marquette University.

Driessen, W. (2001). Anaerobic effluent treatment as an integral part of industrial processes. *Anaerobie 2001 Conference.* Klatovy, Czech Republic.

Driessen, W., & Vereijken, T. (2003). Recent developmens in biological treatment of brewery effluent. *Proceedings of 9th Brewing Convention.* Victoria Falls.

Eckenfelder, W. W. (2000). *Industrial Water Pollution Control.* McGraw-Hill.

Lettinga, G., & Hulshoff, L. W. (1991). UASB Process design for various types of wastewaters. *IAWPRC - Water Science & Technology* , 87-107.

Passeggi, M. (2009). Integrated anaerobic treatment of dairy industrial wastewater and sludge. *IWA Water Science and Technology* , 501 - 506.

Ross, W. R. (1989, Oct.). Anaerobic treatment of industrial effluents in South Africa. *Water SA* .

Sindt, G. (2006). Environmental Issues in the Rendering Industry. In N. R. Association, *Essential Rendering* (pp. 245 - 256).

Speece, R. (2008). *Anaerobic Biotechnology and Odor/Corrosion Control for Municipalities and Industries.* Archae Press.

Totzke. (2009). Anaerobic treatment technology overview. *Anaerobic Treatment of High-Strength Industrial Waste.* Milwaukee: Marquette University.

Van Lier, J. B. (2008). High-rate Anaerobic Wastewater Treatment. *IWA - Water Science and Technology* , 1137 - 1148.

(108b) Comparing Multiple Oxidizers Ability to Replace a Chlorine Pre-Oxidizer Addition In a Drinking Water Filter Plant

Jeff Cochran, Patty Barron and April Nabors, Engineering, Birmingham Water Works Board, Birmingham, AL

Plant optimization testing with a state of the art mobile Pilot Plant was conducted to reduce Disinfection By-Product (DBP) formation from the reaction of Chlorine and DBP precursors in the drinking water filter plants. One approach for reducing DBP formation within a utility is maximizing the reduction of Total Organic Carbon (TOC). Unfortunately, some alternatives such as GAC and/or enhanced coagulation, which can have a profound effect on TOC removal, are not always the optimized options.

While Chlorine residuals are required in the distribution system for drinking water, minimizing the contact time between DBP precursors and Chlorine reduces the amount of DBP's. Reducing contact time led to investigations of replacing any Chlorine feed present before the final disinfection dose after the filters. The Birmingham Water Works Board (BWWB) Shades Mountain Filter Plant (SMFP) used a Chlorine Pre-Oxidant addition at the rapid mix to improve filter performance. Since the plant was already optimized through enhanced coagulation for TOC removal a comprehensive Chlorine Pre-Oxidant replacement study was conducted.

The testing took place in the BWWB mobile pilot plant. The pilot plant contained two identical treatment trains that each contained a raw-water skid, a floc/sed skid, and a filter skid. The potential Chlorine Pre-Oxidant candidates were Hydrogen Peroxide, Chlorine Dioxide, and Potassium Permanganate. They were all evaluated against the current use of Chlorine as a pre-oxidant. Parameters such as DBP formation, pH stability, effect on TOC levels, settled and finished turbidity, metals residuals, algal growth, filter head loss, and filter run times were evaluated. Testing was conducted for one year in combination with various coagulants and was subjected to the same source water variations that were present at SMFP because of its testing within the mobile pilot plant.

Each one of the before mentioned oxidizers has an advantage to its use and at least one disadvantage to its use. This required testing several doses and sometimes combinations of oxidizers to minimize each oxidizer's disadvantage while maximizing its advantages.

See more of this Session: Advances in Water Remediation and Treatment Systems
See more of this Group/Topical: Environmental Division

INSIGHT INTO THE APPLICATION OF ADVANCED OXIDATIVE METHOD FOR BIOSORBENT MODIFICATION IN WATER TREATMENT APPLICATION

Dr. Augustine E. Ofomaja, Dr. Bobby E. Naidoo, and Ms. Veronica T. Dladla, *Department of Chemistry, Vaal University of Technology, X021, VanderbiljPark, 1900 South Africa.*

Introduction

Adsorption using waste/dead plant materials as adsorbent is receiving much attention in recent times due to its low value, natural abundance in nature, ability to remove pollutants at low concentrations e.t.c. Numerous research have been conducted to ascertain the potentials and applicability of water and wastewater treatment to eliminate pollutants such as heavy metals, colored compounds such as dyes, and organic pollutants. The results have indicated the promising ability for these materials to be applied as filters to replace or compliment several other expensive and advanced technologies.

Plant materials are usually made up cellulose, hemi-cellulose, lignin extractives, lipids, water, proteins, simple sugars, hydrocarbons, and starch along with other organic components (Ofomaja 2009). In their use as adsorbents, some plants materials are found to cause discoloration of the treated wastewater leading to increased chemical oxygen demand (COD), biological oxygen demand (BOD) and total organic carbon (TOC) levels (Gaballah et al., 1997). Several methods have been applied in the activation of these biomaterials to prevent leaching of colored organic pigments, some of which are treatment with base solutions (sodium hydroxide, calcium hydroxide, sodium carbonate) mineral and organic acid solutions (hydrochloric acid, nitric acid, sulfuric acid), have been reported in literature.

Oxidative methods have been applied in the removal of organic pollutants from aqueous solution and soils with success, these methods include; Wet oxidation and Catalytic wet oxidation (Deepak et al., 2002), Ozone based oxidation (Bila et al., 2005), Electrochemical oxidation (Deng and Englehardt, 2007) and Fenton's oxidation (San Sebastian et al., 2003). These advanced oxidative method as there are called are effective due to the role of a highly reactive radical intermediate such as hydroxyl radical (•OH) as an oxidant. With an oxidation potential (E^0) of 2.80 mV, the •OH radical can rapidly degrade recalcitrant organics such as aromatic, chlorinated, and phenolic compounds (Parsons and Williams, 2004). Once a reaction of the free radical is initiated by ozone or H_2O_2, a series of oxidation reactions occurs in the solution and the radicals rapidly react with most of the target compounds.

It is well known that the hydroxyl radical (•OH) can be generated by the addition of hydrogen peroxide to an aqueous solution of excess ferrous ions under acid conditions. When this mixture is done in the presence of an organic substrate, a complex redox reaction occurs which has been described by Ahn et al. (1999) as follows:

$$Fe^{2+} + H_2O_2 \rightarrow Fe^{3+} + OH^- + {}^{\cdot}OH \quad (1)$$

$$RH + {}^{\cdot}OH \rightarrow H_2O + R^{\cdot} \quad (2)$$

$$R^{\cdot} + H_2O_2 \rightarrow ROH + {}^{\cdot}OH \quad (3)$$

$$R^{\cdot} + O_2 \rightarrow ROO^{\cdot} \tag{4}$$

The Fenton reaction causes the dissociation of the oxidant and the formation of highly reactive hydroxyl radicals that attack and destroy the organic components (Neyens and Baeyens, 2003).

The major advantages of the Fentons oxidation process includes:

(i) Hydrogen peroxide and ferrous ions are usually more stable in a strong acid.

(ii) Fentons reagent has short reaction time among all advanced oxidation processes

(iii) Ferrous iron and H_2O_2 are cheap and non-toxic

(iv) There are no mass transfer limitations due to its homogenous catalytic nature

(v) There is no energy involved as catalyst and the process is easily to run and control.

The various studies on Fenton's reagent modification of biosorbents reported in literature so far have been concentrated basically on measuring optimum modification on the basis of metal ion removal efficiency. No attention so far has been given to changes in surface properties as a result of the variation in parameters such as the Fe^{2+}/H_2O_2 ratio, solution pH, contact time and oxidation reduction potential (ORP). Therefore in this study, insight into changes in surface properties of pine cone powder during its modification using Fentons reagent was examined. Instrumental methods such as scanning electron microscopy (SEM) and Fourier transform infrared spectrometer (FTIR) were applied in the characterization of the untreated and treated pine cone surface. Other parameters of the treated and untreated pine cone including acid number, bulk density and surface negative charge were also determined to monitor the extent of surface modification. Finally, the treated and untreated pine cone samples were applied for the removal of lead from aqueous solution in a batch reactor.

Methods

Sample collection and preparation

A number of pine cones will be collected from a plantation in Sasolburg, Free State, South Africa and transported to the Vaal University of Technology chemistry laboratory. The pine cones will be washed to remove impurities such as sand and leaves. The washed cones will then be dried at 90°C for 48 hr in the oven. The scales on the cones will be removed and blended in a food processing blender. The pine cone powder will then be sieved and particles between 300 μm will be collected and used for analysis.

Fenton's reagent activation

An accurately weighed and oven dried 20 g of cone material will be poured in 500 ml conical flask containing 250 ml Fenton's reagent solution and stirred with 200 rpm for all activation experiment at room temperature. The pine cone was separated from the solution by filtration, washed and dried at 90 °C.

Bulk Density

Bulk density measurement was carried out in a 25 cm^3 density bottle. The dry pine cone powder was added to the density bottle with gentle tapping to ensure that the particles settle to the bottom and all air spaces are filled. The mass of the density bottle containing the pine cone powder was then determined. The mass of pine cone powder that occupied 25 cm^3 was then obtained from the mass of the bottle and pine cone powder minus the mass of empty bottle.

Acid Number Determination

The acid number (AN) before and after the modification process will be determined using the method suggested by Matsuda(1987) .Approximately 0.3 g of pine cone powder will be placed in a flat –bottom flask of 200 ml volume and 10 ml of 0.1 M aqueous solution of HCl and 100 ml of distilled water will be added. The mixture will then be titrated using 0.1 M aqueous solution of HCl and 100 ml of distilled water will be added. The mixture will then be titrated using 0.1 M aqueous solution of KOH in the presence of phenolphthalein as an indicator. The acid number will be determined from the following Eq. (5):

$$\text{LK} = \frac{(V - H).N.56.1}{m}[mgKOH/g] \qquad (5)$$

Where: V-volume of 0.1 M KOH used for the titration of the solution [ml},H-volume of 0.1 M KOH used for the neutralization of 10 ml aqueous solution of 0.1 M HCl [ml], N-concentration of the KOH solution, m-sample weight (g).

Results and discussion

Optimization of H_2O_2/Fe^{2+} ratio

Fenton's activation involves the interaction of hydroxyl radicals (˙OH) produced by the mixture of H_2O_2 and Fe^{2+} with organic materials on the adsorbent surface. The interaction produces complex redox reaction steps (Ahn et al., 1999) that leads to the attack and destruction organic components of the pine material. This interaction has been shown to depend on the H_2O_2/Fe^{2+} ratio, solution pH and time to contact. To monitor the progress of Fenton's activation, Argun et al. (2008) monitored the ORP of the solution and found that it correlates well with the activation of pine cone surface. Walling (1975) simplified the overall Fenton chemistry to occur via consumption of H^+ ions from aqueous solution (Eq. 6), that is, as hydroxyl radicals are generated and reacted H^+ ions are removed from solution and solution pH will increase.

$$2Fe^{2+} + H_2O_2 + 2H^+ \rightarrow 2Fe^{3+} + 2H_2O \qquad (6)$$

In our investigation, we choose to determine the optimum of H_2O_2/Fe^{2+} ratio by keeping the Fe^{2+} concentration constant and varying the H_2O_2 concentration in one experiment for 30 min, and varying the Fe^{2+} concentration at the optimized concentration of H_2O_2 to optimize the Fe^{2+} concentration. The initial pH of solution was fixed between 3.0 and 4.5 which have been proven to be optimum for Fenton's reaction from literature. The progress of the activation was monitored by measuring the ORP ratio and the change in H^+ ion concentration at the end of the activation process. Figs. 1a and b shows results obtained for the optimization process. When constant Fe^{2+} concentration of 500 mg/L and varying H_2O_2 concentrations (140,000 to 1,000 mg/L) was used at initial solution pH ranging between 2.4 and 2.7, the final solution pH's were seen to increase slightly above the initial solution pH's. The change in H^+ ions concentration was measured as follows:

$$\Delta[H^+] = \Delta[H^+_{\text{Initial}}] - \Delta[H^+_{\text{final}}] \qquad (7)$$

And $$pH = -\log [H^+] \qquad (8)$$

The values of change in hydrogen ion concentration (ΔH^+) was observed to increase with decreasing H_2O_2 concentration from 140,000 (1.648 mmol/L) and drops after reaching a maximum at 100,000 mg/L (2.006 mmol/L). Increasing values of ΔH^+, implies that more H^+ ions are consumed from solution while reducing values implies that less H^+ ions are consumed. These results indicate that 100,000 mg/L of H_2O_2 gave the highest consumption of H^+ ions which is a H_2O_2/Fe^{2+} ratio of 200. The trend was also found to be similar to the ORP measured for the different solutions. The ORP increased with decreasing H_2O_2 concentration from 140,000 (196.3 mV) and drops after reaching a maximum at 100,000 mg/L (218.5 mV). Argun et al. (2008) showed that optimum activation conditions coincided with optimum solution ORP measured. This suggests that ΔH^+ of the solution can be used to determine the optimum activation conditions.

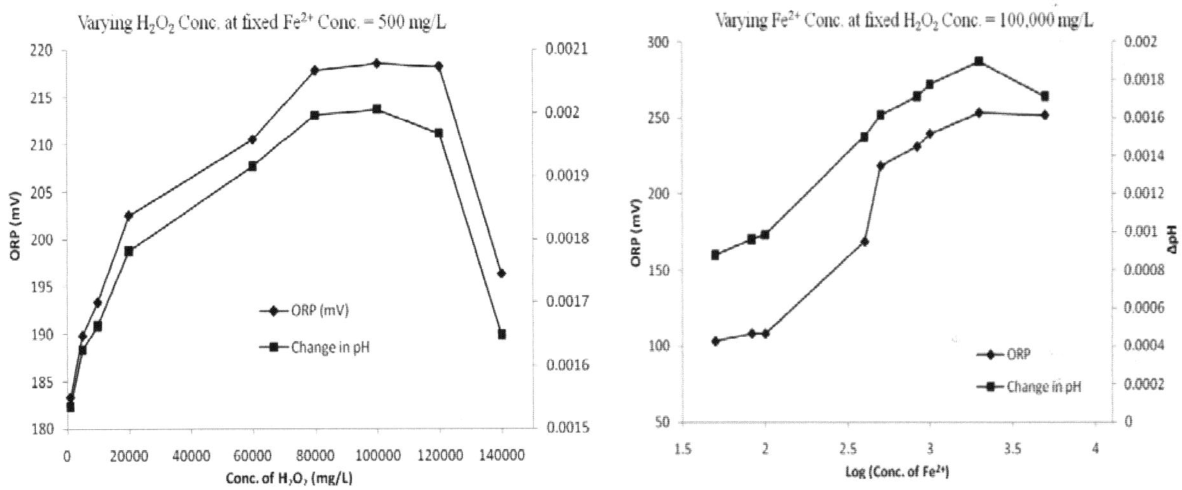

Fig. 1: Effect of varying (a) H_2O_2 on a fixed amount of $Fe^{2+}$and (b)Fe^{2+} at fixed H_2O_2.

When the H_2O_2 concentrations were kept constant at 100,000 mg/L on the other hand and the concentration of Fe^{2+} varied between 5000 and 50 mg/L, ΔH^+ of the solution increased from 1.714 to 1.776 mmol/L as Fe^{2+} reduced from 5000 to 1000 mg/L. As Fe^{2+} concentration decreased below 1000 mg/L, the ΔH^+ of solution reduced to 0.880 mmol/L (Fe^{2+} = 50 mg/L). The trend in the relationship between ORP and Fe^{2+} concentration were also similar, with a maximum ORP values occurring at 2000 mg/L (253.5 mV) of Fe^{2+}. The ratio of concentrations of H_2O_2 to Fe^{2+} in this case was 100 which gave a higher ΔH^+ and ORP than for H_2O_2 to Fe^{2+} ratio of 200 as obtained in the first experiment. It can therefore be concluded that the concentrations of H_2O_2 and Fe^{2+} applied are as important to the activation process as the ratios of H_2O_2 to Fe^{2+}, since the ratio $H_2O_2/Fe^{2+} = 120$, from the first experiment where 500 mg/L of Fe^{2+} was used gave much lower ΔH^+ and ORP (1.915 mmol/L and 210.5 mV).

Optimization of Contact time

The effect of contact time on the activation reaction was also examined. The activation reaction was performed under similar conditions but the contact time was varied between 15 to 120 min. The results for the experiments are reported in Table 2 below. The effect of contact time on the activation reaction was also examined. The activation reaction was performed under similar conditions but the contact time was varied between 15 to 120 min. The results for the experiments are reported in Table 3.2 below.

Table 2: Variation in contact time of Fenton's activation using 100,000 mg/L H_2O_2 and 1000 mg/L Fe^{2+}.

Contact time (min)	Initial pH	Final pH	ΔH^+ ion (mmol/L)	ORP (mV)
15	2.42	2.69	1.776	251.9
30	2.42	2.72	1.896	253.5
45	2.42	2.71	1.852	239.6
60	2.42	2.68	1.713	231.1
120	2.42	2.48	1.491	218.5

The results revealed that increasing contact time of activation from 15 to 30 min increased the ΔH^+ and ORP from 1.776 mmol/L and 251.9 mV to 1.776 mmol/L and 251.9 mV. Further increase in contact time caused both the ΔH^+ and ORP to decrease. Therefore, optimum contact time for the activation solution and 20 g of pine cone powder is 30 min.

FTIR Analysis

Since the pine cone surface consist of several functional groups, analysis of these groups before and after treatment was performed in other to determine the after of Fenton's activation on the pine cone surface. The RTIR analysis of raw pine cone. Several peaks were observed from the spectra indicating that pine cone is composed of various functional groups. The broad intense spectra bands observed at 3342.65 cm^{-1} are indicative of unbounded–OH and the peak observed at 2927.95 cm^{-1} represents the aliphatic C–H group. The peak at 1607.67cm^{-1} corresponds to the C=O stretch. The peaks between 1023.26 and 559.32cm^{-1} may be assigned to the –C–C– and –CN stretching, respectively. A comparison between the FTIR spectra's of the raw pine cone and the pine cone modified with $H_2O_2/Fe^{2+} = 200$ and $H_2O_2/Fe^{2+} = 100$ (samples with the better activation) showed some differences indicating the functional groups on the surface has been modified. The band at 3342.65 cm^{-1} and peak at 1697.57 cm^{-1} increased and the wavelength slightly shifted for the modified samples. This may suggest that the C=O and –OH groups increased of the surface after activation. An increase in carboxylic functional group is only likely to occur if Fenton's activation had led to oxidation of primary or secondary alcohols of the pine cone surface.

SEM analysis

The SEM micrograph of the untreated pine cone powder along with the samples treated with H_2O_2/Fe^{2+} ratios of 200 and 100 (not shown) indicates the surface of the material did not under any major changes except for the appearance of pores spaces.

Acid number analysis

Pine cone contains 65.3 % of cellulose (hemi-cellulose and α-cellulose) which is composed largely of hydroxyl groups (-OH). The reaction of –OH groups on wood products have been measured by acid number (AN mg KOH/g) determination. As the number of –OH groups decreases the acid number increases. Fig. 2a shows the acid number values for pine cone treated with H_2O_2 of concentration ranging from 140,000 to 1000 mg/L with a fixed concentration of Fe^{2+} (500 mg/L).

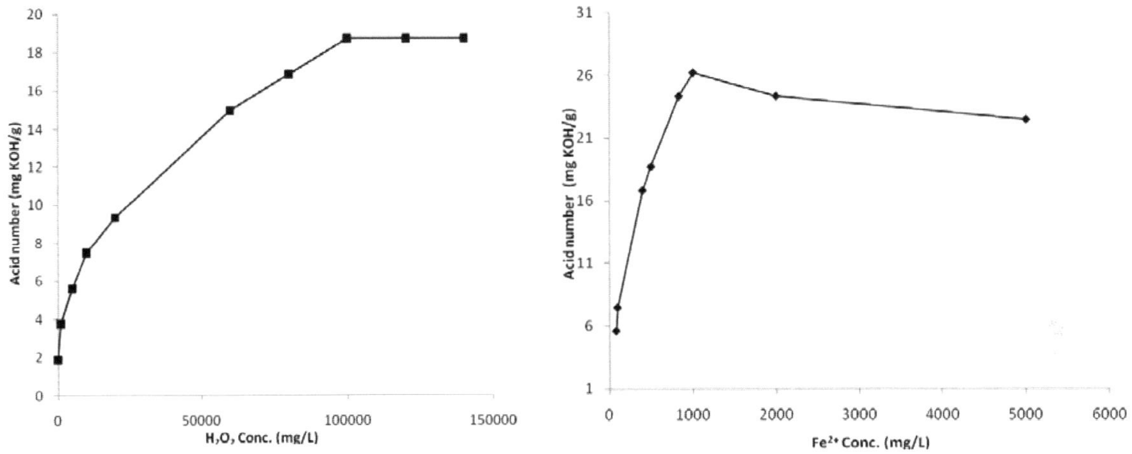

Fig. 2: Relationship between acid number of activated pine cone and (a) H_2O_2 concentration at fixed Fe^{2+} concentration of 500 mg/L and (b) Fe^{2+} concentration at fixed H_2O_2 concentration of 100,000 mg/L.

The plot reveals that the raw pine cone had the least acid number (1.87 mg KOH/g), indicating that it had the highest amount of –OH groups on its surface. Doczekalska et al. (2007) obtained acid number for pine 1.1 mg KOH/g which is close to our result. On treatment of the pine with Fenton's reagent of 1000 mg/L H_2O_2 and 500 mg/L Fe^{2+}, the acid number increased to 3.74 mg KOH/g, indicating that –OH groups have been converted or used up. Increasing the concentration of H_2O_2 up to 100,000 mg/L increased the acid number to 18.7 mg KOH/g. It was observed that further increase in the concentration of H_2O_2 above 100,000 mg/L did not produce any incensement acid number. H_2O_2 concentration of 100,000 mg/L gave the optimum modification in this experiment and also coincided with the concentration of H_2O_2 at which the maximum amount of –OH groups were converted. When the H_2O_2 concentration was kept constant at 100,000 mg/L and Fe^{2+} concentration varied between 50 and 5000 mg/L for the pine cone activation, the acid number of all the activated samples were higher than for the raw pine cone (Fig. 3.2b). The acid number was found to increase (3.74 to 26.18 mg KOH/g) as the concentration of Fe^{2+} increased from 50 to 1000 mg/L. When Fe^{2+} concentration was increased above 1000 mg/L to 5000 mg/L the acid number dropped to 22.44 mg KOH/g. This concentration of Fe^{2+} also corresponds to the concentration of Fe^{2+} that gave the best activation in the second experiment. It will be noticed that the values of acid number produced by H_2O_2/Fe^{2+} ratio of 100,000/1000 gave a higher acid number (22.44 mg KOH/g) than for the H_2O_2/Fe^{2+} ratio of 100,000/500 (18.7 mg KOH/g). This means that a larger number of –OH groups are converted in the former case than in the latter.

As have been suggested, the reduction of –OH groups on the pine cone surface may have been due to oxidation by Fenton's reagent. According to our results, -OH groups were found to have reduced in amount and carboxylic acid groups were found to have increased. Moody (1963) showed that D-glucose and D-fructose in the presence of Fenton's reagent can be converted to glycollic, glyoxylic, oxalic and trihydroxybutyric acids. Collinson and Thielemans (2010) suggested that the initial step of the reaction is the selective oxidation of the –OH group on carbon 6 forming a carboxylic acid. The presence of Fe^{3+} in the solution catalyses further oxidation and a carboxylic acid and an aldehyde is produced. In effect two hydroxyl groups of alcohol are consumed to produce a carboxyl and an aldehyde group. This may likely account for the reduction in the amount of –OH groups as predicted by the acid number determination and the increase in carboxylic acid functional groups as seen in the increase of the band and peak at 3342.65 cm^{-1} and peak at 1697.57 cm^{-1} observed for the activated pine cone samples as compared with the raw sample.

Lead ion adsorption

Results on the lead ion adsorption by untreated and Fenton's reagent treated pine cone reveals that Fenton's reagent significantly increases lead ion uptake and that treatment with 200,000 mg/L H_2O_2 and 1000 mg/L Fe^{2+} (23.32 mg/g) produced better adsorption than treatment with 100,000 mg/L H_2O_2 and 500 mg/L Fe^{2+} (19.12 mg/g) and the untreated (9.34 mg/g).

Conclusion

Our study, it is suggested that Fenton activation did not only destroy the organic components of pine cone and increased porosity, but oxidized the material surface producing additional acidic functional groups as was observed from the increasing values of acid number and the functional group determination. Varying the H_2O_2/Fe^{2+} increased the pH change and ORP to a maximum values at H_2O_2/Fe^{2+} = 100, and after this value these parameters were found to decrease. Therefore, maximum lead(II) removal was due to increased acid functional groups on the pine cone surface rather than increased porosity.

Reference

1. Ofomaja, Augustine E. (2009), "Removal of transition metals from industrial wastewater using natural biosorbents". Doctoral Thesis submitted to Vaal University of Technology.

2. Gaballah, I., Goy, D., Allain, E., Kilbertus, G., Thauront, J. (1997), "Recovery of copper through decontamination of synthetic solutions using modified barks". Met. Metall. Trans. B 28 pp. 13–23.

3. Deepak, B.A., Suresh K.B., Irfan S., Jaidev P. (2002), "Catalytic wet oxidation: an environmental solution for organic pollutant removal from paper and pulp industrial waste liquor". Applied Catalysis A: 236 pp. 255-262.

4. Bila, D.M., Montalvao, A.F., Silva, A.C., Dezotti, M. (2005) "Ozonation of a landfill leachate: evaluation of toxicity removal and biodegradability improvement". *Journal of Hazardous Materials* 117 pp. 235–242.

5. Deng, Y., Englehardt, J.D., (2007),, "Electrochemical oxidation for landfill leachate treatment". *Waste Management*, 27 pp.380–388

6. San, Sebastian N., Fernandez, J.F., Segura, X.F., Ferrer, A.S., (2003), "Pre-oxidation of an extremely polluted industrial wastewater by the Fenton's reagent". *Journal of Hazardous Materials* 101 pp. 315–322.

7. Ahn, D.H., Chang, W.S., Yoon, T.I., (1999), "Dyestuff wastewater treatment using chemical oxidation, physical adsorption and fixed bed biofilm process". Process Biochem. 34 pp. 429–439.

8. Neyens, E., Baeyens, J., (2003), A review of classic Fenton's peroxidation as an advanced oxidation technique. J. Hazard. Mater. B 98 pp. 33–50.

9. Argun, M.E., Dursun, S. (2008), "A new approach to modification of natural adsorbent for heavy metal adsorption". *Bioresource Technology* 99 pp. 2516-2527.

10. Walling, C., Kato, S. (1971), "The oxidation of alcohols by Fenton's reagent: the effect of copper ion". J. Am. Chem. Soc. 93 (1971) 4275–4281.

11. Moody, G.J. (1964) The actions of hydrogen peroxide on carbohydrates and related compound. Advances. in Carbohydr. Chem. 19:149–179

12. Collinson, S.R., and Thielemans W. (2010) "The catalytic oxidation of biomass to new materials focusing on starch, cellulose and lignin" Coordination Chemistry Reviews 254 pp. 1854-1870.

Photocatalytic Degradation of Organic Contaminants in Wastewater Using Visible Light Irradiation

Muhammad A. Daous, Ahmed Arafat[*] and Esam Al-Shareef
Chemical & Materials Engineering Department, King Abdulaziz University, P.O. Box 80204, Jeddah, 21589, Saudi Arabia, Tel: +9662-695-1247, Fax: +9662-6952257; Email: *mdaous@kau.edu.sa*

Summary

Titanium dioxide nanoparticles successfully anchored into 3-D mesoporous material (TUD-1) using a sol-gel preparation strategy were prepared and tested as photocatalysts, employing only visible light (λ = 447 nm), for the treatment of wastewaters containing organic contaminants, such as phenol and azo dyes. Samples of different Si/Ti ratio, ∞, 100, 75, 50 and 25 and 10 *w/w* were prepared and characterized by X-ray diffraction (XRD), transmission electron microscopy (TEM), ultraviolet spectroscopy (UV), and nitrogen adsorption at 77K. These samples were found to have mesoporous structure and only samples of high loadings of titania were found to have anatase in very small amount, while low loading samples didn't show separate phases of TiO_2. Diffuse reflectance UV-VIS spectra of the catalysts showed absorption bands in both the UV and visible region of the spectra. The prepared samples were tested for the photcatalytic treatment of simulated wastewaters contaminated with phenol and azo dye as organic contaminants by irradiation with visible light only (λ = 447 nm).

The newly developed Ti-TUD-1 photocatalysts showed excellent activity towards the removal of the tested organic compounds. Up to 99% of phenol and 98% of azodyes were successfully removed from the contaminated wastewater. Moreover, the catalysts showed remarkable stability under the relatively mild irradiation conditions used. Repeated photodegradation tests demonstrated remarkable reproducibility and showed no decrease in photocatalytic activity after 5 cycles.

Introduction

Wastewater effluent recycling and Reuse is of utmost importance for a worldwide sustainable water management portfolio that aims to minimize water shortage in water scarce environments [1,2]. The frequent occurrence of pesticides and phenolic compounds in wastewater and associated environments has heightened concerns over public health due to the high toxicity and bio-recalcitrant nature of such contaminants [3]. Consequently, considerable efforts have been devoted to develop suitable purification methods that can effectively destroy these bio-recalcitrant organic contaminants. However, due to their incomplete removal during currently used wastewater treatment methods, they are ubiquitous in secondary wastewater effluents at low concentration.

Heterogeneous photocatalytic oxidation (HPO) processes employing catalysts such as TiO_2, ZnO, etc., using UV light has shown promising results for the degradation of persistent

[*] Permanent address: Chemistry Department, Faculty of Science, Helwan University, Ain Helwan, Helwan, 17790, Egypt. Tel: +20225552467, Fax: +20225552468, Email: akhamis@kau.edu.sa

organic pollutants, and their conversion to more biologically degradable and less toxic substances [4-6]. However, a large number of organic contaminants cannot be fully degraded by such methods. In order to meet increasingly more stringent environmental regulations and laws, more efficient processes for the removal of organic pollutants are needed. Among the most promising and advanced methods being investigated, photocatalysis at semiconductors and at TiO_2, in particular, receives a considerable attention as reflected by the recent large body of research published in this field. Titanium dioxide, however, can be easily degraded when subjected to the relatively harsh UV irradiation. Moreover, use of UV irradiation is not attractive in a large industrial scale water treatment processes because of its potential harmful effects on humans. Both of the problems cited above could be resolved if UV irradiation could be replaced by a harmless visible light irradiation.

In this work, TiO_2 nanoparticles are anchored in 3-D mesoporous matrix of TUD-1 material. The downsizing of TiO_2 particles to the nanometer scale leads to a significant reduction of their band gap, as well as a large increase in the surface area of the produced phtocatalyst. This allows for a better performance of TiO_2 photocatalysts in ridding wastewater of organic contaminants in a much safer process.

Experimental

Pure siliceous TUD-1 and five samples of Ti-TUD-1 with different Si/Ti ratio of (100, 75, 50, 25 and 10) were prepared according to the procedures described by Shan *et al.* [7]. These samples were coded as TUD-1, Ti-100, Ti-75, Ti-50, Ti-25 and Ti-10. A homogeneous synthesis mixture was prepared by mixing of titanium tert-butoxide, tetraethylorthosilicate (TEOS) and triethanol amine (TEA). Typically, TEA (97% ACROS) was added dropwise into a mixture of titanium (IV)butoxide (99% ACROS), and TEOS (ACROS) while stirring. After stirring for about 0.5 h, water was added dropwise into the above mixture, followed by the addition of tetraethylammonium hydroxide (TEAOH, 25% in water, Aldrich) to control the hydrolysis and polycondnesation of the alkoxides while stirring. Finally clear synthesis mixture was obtained with a molar oxide ratio composition of SiO_2:0.02TiO_2:0.3TEAOH:2TEA:11H_2O. The mixture was aged at room temperature for at least 24 h, dried at about 100°C and then calcined in air at 750°C for 10 h with a ramp rate of 1°C/min.

XRD patterns were recorded using Cu Kα radiation (λ = 1541 nm) on a Philips PW 1840 diffractometer equipped with a graphite monochromator. The samples were scanned over the range of 0.1–80° 2θ with steps of 0.02°.

Nitrogen sorption isotherms were recorded on a Quantachrome NOVA 2200e at 77 K. Mesoporosity was calculated from the adsorption branch using the Barret–Joyner–Halaenda (BJH) model. The BET method was used to calculate the surface area (S_{BET}) of the samples, while mesopores volume (V_{meso}) was determined with the t-plot method according to Lippens and Boer. Diffuse reflectance UV–vis spectra were recorded on a CaryWin 300 spectrometer under atmospheric conditions using $BaSO_4$ as reference. Samples were scanned from 190 to 800 nm. The UV-Vis absorption data were converted to Kubelka-Munk units using the following formula:

$$F(R) = \frac{(1-R)^2}{2R}$$

Transmission electron microscopy (TEM) was performed using a Philips CM30T electron microscope with LaB6 filament as the source of electrons operated at 300 KV.

Photocatalytic Experiments

Photocatalytic experiments were performed on a Heraeus, photoreactor system 1 with an implemented magnet-driven circulation pump, which provided a complete mixing of the reaction liquid. Irradiation was effected by means of a low pressure mercury lamp TQ718 using 700 W power supply. Radiation wavelength was adjusted by variation of the radiation flux Φ, to the required value. The low pressure mercury lamp could be operated in the wavelength range of 240 to 577 nm.

Photocatalytic experiments were performed using 250 ml samples of simulated wastewater contaminated with either phenol or peach red azo dye. The reaction was followed by sampling 3 ml of the reaction mixture withdrawn at different time intervals after ceasing stirring for one minute to allow most of the catalyst particles to settle down, while maintaining irradiation. Phenol concentration in these samples was measured by a Shimatzu UV-vis spectrometer, UV-1800.

Results and discussions

The X-ray diffraction patterns of the calcined Ti-TUD-1 samples show an intense peak at 0.8-2.0° 2θ, corresponding to *d*-spacing of about 8.6 nm. This peak is identical to mesostructured materials, as reflected by Fig. 1.

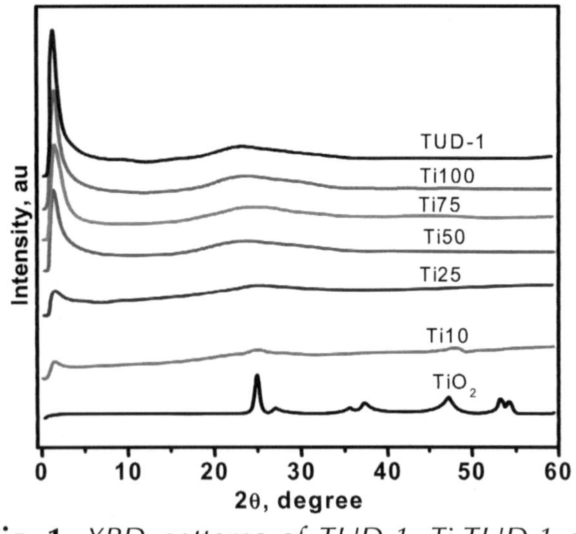

Fig. 1. *XRD patterns of TUD-1, Ti-TUD-1 and TiO_2 calcined samples at 550 °C.*

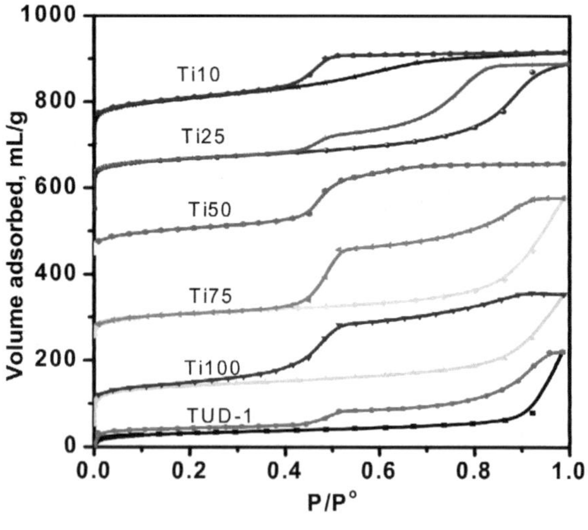

Fig. 2. *Nitrogen adsorption/desorption isotherms at 77 K for TUD-1 sample and Ti-loaded TUD-1 samples.*

Typical nitrogen adsorption isotherms of the prepared samples are shown in Figure 2. The obvious hysteresis loops can be found at P/P° values of 0.45 to 0.8 which reflect a maximum peak of pore size distribution around 5-8 nm. A plateau at the relative pressure values of 0.8-1.0 indicates the presence of a textural mesoporosoity. Moreover, the specific surface area for these samples reaches 400 m^2/g, while the pore volume reaches 0.89 cm^3/g. The wall thickness is estimated to be around 3.6 nm, which is thicker than MCM-41.

TEM images of siliceous TUD-1 showed a sponge-like structure as shown in figure 3a. That was also the case for Ti-TUD-1 with small titanium loading (i.e. Si/Ti ratio > 50), figure 3b. This indicates that at these relatively low titanium loading levels, all Ti is incorporated in the mesoporous framework. On the other hand, at higher levels of titanium loading (i.e. Si/Ti ratio < 50), nano-particles of TiO_2 were detected, figure 3c. The size of the nano-particles corresponds to the pore size of the matrix and indeed the number of the visible nano-particles per image depends on the titanium loading. Moreover, samples of Ti-10 were found to have small amounts of anatase.

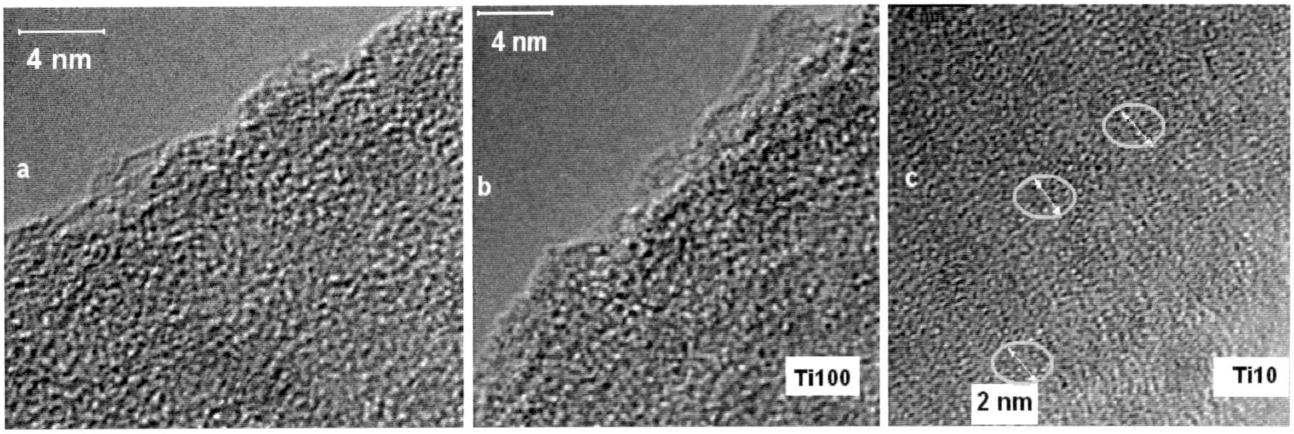

Fig. 3. *TEM images for (a)TUD-1, (b) Ti-100 catalyst sample and (c) Ti-10 catalyst sample, green circles mark anatase clusters imbedded in the mesoporous materials.*

The UV-VIS spectra of figure. 4 show a narrow band centered at 225 nm, which is attributed to the charge-transfer transition associated with isolated Ti(IV) framework sites in tetrahedral co-ordination [8]. The other peak in the 320-350 nm range is an indication of the presence of poly titanium (Ti-O-Ti)n anatase clusters [9,10]. The intensity of this peak increases with titanium loading, which is consistent with the formation of a crystalline TiO_2 phase along with a titanosilicate phase. However, the size of these crystals is beyond the detection limit of the XRD. It is known that pure TiO_2 displays no significant absorbance in the visible light region due to its large band gap (3.2 eV).

The calcined Ti-loaded TUD-1 samples also display some absorption in the 400-700 nm, which indicates that these samples have photocatalytic activities in this region. The absorbance in this region is more pronounced in highly Ti-loaded samples. The reason for this absorption behavior is not fully understood because the TiO_2 clusters are located in a rather complex environment of silica and titanosilicate.

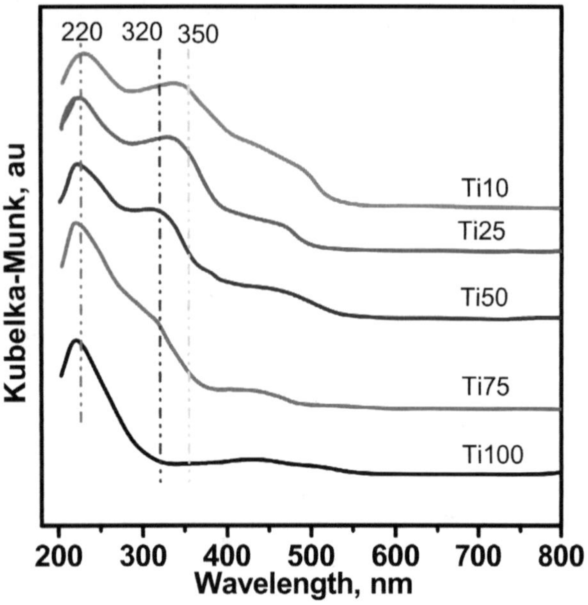

Fig. 4. *UV-VIS spectra of titanium-loaded TUD-1.*

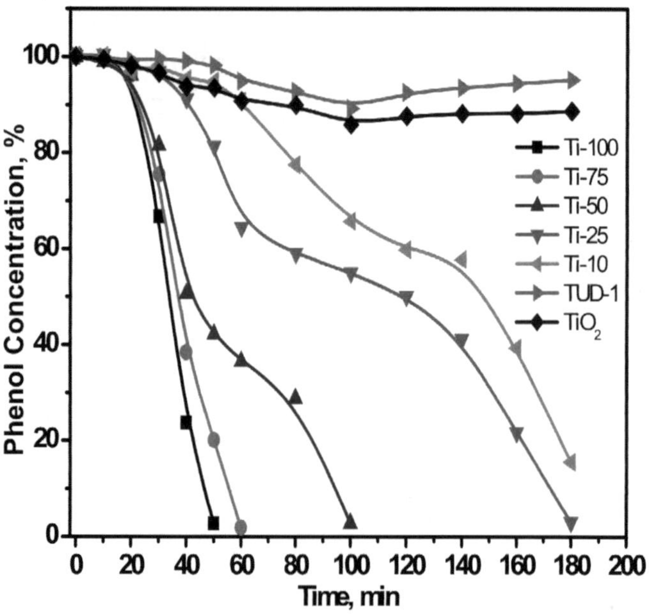

Fig. 5. *Photocatalytic degradation of 250 ml, 500 ppm phenol solution containing 0.5 g of various phtocatalysts as a function of 447 nm irradiation-time.*

Photocatalytic degradation of phenol on Ti-loaded TUD-1 photocatalysts

Preliminary experiments showed that upon 20 h of 447 nm irradiation of a sample of phenol solution of 500 ppm initial concentration, (pH 6.8), no phenol could be detected based on UV-Vis measurements. Moreover, no phenol was detected on the filtered solid catalyst as well, most likely due to electrostatic repulsion. Nonetheless, phenol degradation results suggest that the photocatalytic reactions initially starts with adsorption of phenolic species on the catalyst surface, which explains the slight increase in the detectable phenol in the first 30 minutes of irradiation, as shown in figure 5

Figure 5 also compares the degradation of 250 ml of 500 ppm phenol solution in water using 0.5 g of the various photocatalysts tested in this study. No apparent degradation of phenol takes place when pure TUD-1 material is used as a catalyst, due to the absence of photocatalytic active sites. Similarly, TiO_2 is also photoctalytically inert in the visible light region. On the other hand, titanium-loaded TUD-1 mesoporous catalysts show significant photocatalytic activity in the phenol degradation reaction. Moreover, Increasing titanium loading enhances the degradation of phenol substantially. Ti-10 and Ti-25 samples are the most active, achieving 100% phenol degradation in 50 and 60 minutes, respectively. This activity towards phenol degradation can b attributed to the presence of titania naoparticles within the pores of TUD-1

These results clearly show that Ti-loaded mesoporous TUD-1 catalysts exhibit relatively higher photocatalytic activity in phenol degradation compared to other reported titanium-based phtocatalysts in the literature [11-13]. This is a result of a much shorter time needed for the phenol adsorption on the TUD-1 matrix and the lower resistance to diffusion offered by the spongy nature of the TUD-1 structure.

Photodegradation of peach red azo dye

The structure of the peach red azo dye tested in this work is given in Scheme 1. To insure dye degradation was solely due to photocatalytic reaction, a 500 ppm aqueous dye solution was irradiated with 447 nm for 5.0 h. No observable degradation of the dye was detected. When 0.5 g of pure calcined TUD-1 catalyst was added to a 500 ppm aqueous dye solution irradiated with 447 nm for 3.0 h, very slight (not more than 4%) degradation of the dye was observed, indicating that TUD-1 is practically inactive photocatalytically.

Degradation percent of the tested peach red azo dye, using the various catalysts tested in this study as a function of the irradiation-time is depicted in figure 6.

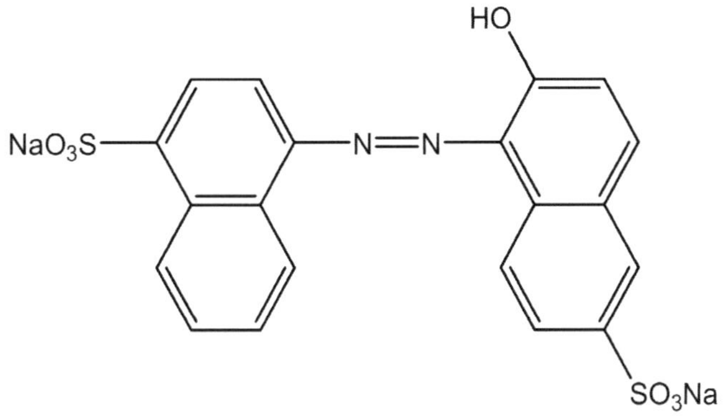

Scheme 1. *Structure of peach red azo dye*

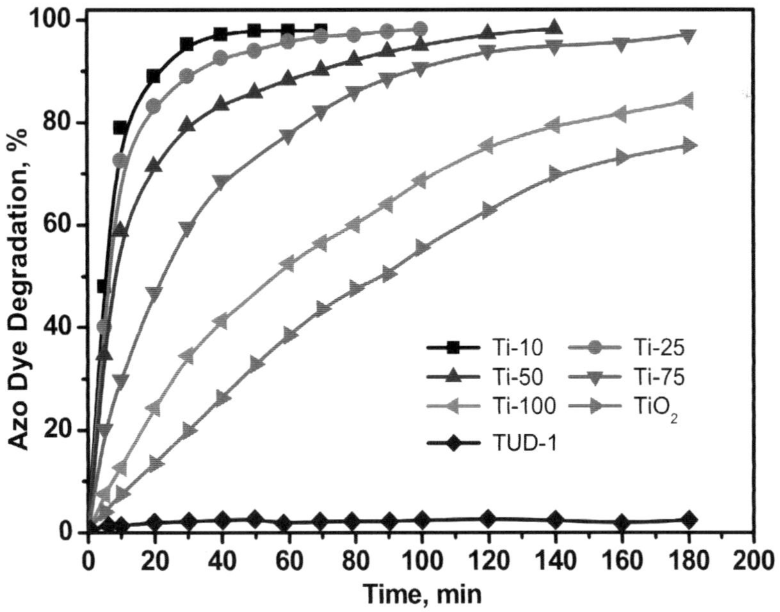

Fig. 6. *Photocatalytic degradation of a 250 ml samples of 500 ppm peach red azo dye solution using 0.5 g of the various phtocatalysts tested as a function of 447 nm irradiation-time.*

Increasing titanium loading enhances the degradation of azo dye significantly. Ti-10 and Ti-25 Samples are the most active photocatalysts, producing close to a 100% degradation of the azo dye in 70 and 100 minutes, respectively. Other lower Ti-load active samples require around 3.0 hours of irradiation time to achieve similar degradation of the azo dye. Ti-75 and Ti-100 sample exhibit much lower degradation activity. These samples do not contain extra framework TiO_2 and most of the titanium is included in the TUD-1 framework. This confirms the necessity of nonframework TiO_2 to achieve high photoactivity.

Stability of the photpcatalysts

Catalysts used in the degradation of phenol and azo dye experiments were separated at the end of each reaction tests. These catalysts were then reactivated by calcining in air in a procedure similar to their initial preparation, and retested in similar contaminants degradation reactions. No significant change in the photocatalytic activity of the reused catalysts was observed. This procedure was repeated for four more times without any significant change in the percentage of achieved degradation of both contaminants. These results reveal that these catalysts are very stable under the relatively mild and safe experimental conditions employed in thses tests.

Conclusions

Titanium could be successfully incorporated in the 3D mesporus material TUD-1 by a sol gel preparation procedure. Titanium is incorporated in the TUD-1 at relatively low Si/Ti ratios. Non-framework TiO_2 species could be detected by nitrogen adsorption at 77K, TEM and UV-VIS spectra but not with XRD. Ti-loaded TUD-1 catalysts show a significant activity in the visible light region. Catalysts with relatively low Si/Ti ratio were very active in photocatalytic degradation of phenol and peach red azo dye while catalysts with relatively high Si/Ti ratio were less active. The presence of non-framework TiO_2 species enhances the photocatlytic activity of the catalysts.

Acknowledgements

Authors greatly acknowledge King Abdulaziz City for Science and Technology, Saudi Arabia, for financing Project No. 8-WAT-138-3, as part of The National Plan for Strategic Projects.

References

[1] V.G. Mitchell, R.G. Mein, T.A. McMahon, Australian Journal of Water Resources. 6 (2002) 31-43.

[2] J.C. Radcliffe, Desalination. 187 (2006) 77-87.

[3] E. Eriksson, A. Baun, P.S. Mikkelsen, A. Ledin, Desalination. 215 (2007) 187-197.

[4] A. García-Ripoll, A. Arques, R. Vicente, A. Domenech, A.M. Amat, Journal of Solar Energy Engineering, Transactions of the ASME. 130 (2008) 0410111-0410115.

[5] A. García, A.M. Amat, A. Arques, R. Sanchís, W. Gernjak, M.I. Maldonado, I. Oller, S. Malato, Environmental Chemistry Letters. 3 (2006) 169-172.

[6] J.J. Vora, S.K. Chauhan, K.C. Parmar, S.B. Vasava, S. Sharma, L.S. Bhutadiya, E-Journal of Chemistry. 6 (2009) 531-536.

[7] Z. Shan, J.C. Jansen, L. Marchese, T. Maschmeyer, Microporous and Mesoporous Materials. 48 (2001) 181-187.

[8] S. Klein, B.M. Weckhuysen, J.A. Martens, W.F. Maier, P.A. Jacobs, Journal of Catalysis. 163 (1996) 489-491.

[9] A.K.L. Sajjad, S. Shamaila, B. Tian, F. Chen, J. Zhang, Applied Catalysis B: Environmental. 91 (2009) 397-405.

[10] S. Sajjad, S.A.K. Leghari, F. Chen, J. Zhang, Chemistry - A European Journal. 16 (2010) 13795-13804.

[11] C.H. Chiou, C.Y. Wu, R.S. Juang, Chemical Engineering Journal. 139 (2008) 322-329.

[12] W. Zhang, Y. Li, C. Wang, P. Wang, Desalination. 266 (2011) 40-45.

[13] H. Ilyas, I.A. Qazi, W. Asgar, M.A. Awan, Z.U.D. Khan, Journal of Nanomaterials. 2011 (2011).

A study of the Next Generation Water Conditioning Unit to fulfill Iraqi Needs for Drinkable Water

Dr. Hadeer K. Hamid, Water & Air Enterprises Ltd., Vancouver, BC. Canada

The Summary

A powerful effective system, eco-friendly and requires no suspect chemical additives called the "Generator" provided by LVPG International is used in this study. The LVPG Generator uses only water, salt, and electricity to generate two solutions namely; AnoFluid and CathoFluid. AnoFluid, the main solution used in the disinfection of drinking water also known as "Mixed Oxidants" is studied thoroughly in this research to test its applicability as a disinfecting agent in order to replace the use of gaseous chlorine, hence, avoiding unnecessary risks associated with its application as well as the elimination of the danger of by - products associated with its use, such as THM, Haloacetic acid, Haloketones and many others. The disinfection process has been examined and tested on Iraqi water samples and showed positive and promising results. According to the test results, it was found that the AnoFluid is capable of killing bacteria within 1 minute in a controlled environment. The result is clear, clean non-contaminated water with no chemical residues. A simulation model for the LVPG generation process is then set-up using ChemCAD simulation software in order to provide a generalization of the proposed facility, if constructed according to proposed dimensions and operations.

Introduction

One of the major problems in Iraq at the present time is water pollution and the unsuitability of drinking water. Serious efforts have been spent in order to overcome this serious problem. It is required to provide high quality drinking water solution that complies with the world drinking water standards (WHO) [1] and accepted by the water health organization in order to replenish houses, factories, schools and hospitals, etc

Chlorination is the most widely used method for disinfecting water supply in Iraq. It is regarded as a reliable, cost - effective method for disinfecting water for drinking purposes. However, in remote areas, the application of chlorine presents problems where difficulties such as the following may exist.

- Distance from the place of chlorine manufacture in addition to the fact that transport and delivery schedules are unreliable and non-existent in some areas according to the current security problems in many places of Iraq.
- Lack of sufficient expertise on the proper dosing, since the administration of the correct dose of chlorine is of greatest importance, not only for the safety of the supply, but also for the freedom of the treated water from odor and taste.
- As indicated in the literature review [2-8] chlorination can result in the formation of harmful by products such as Trihalomethanes, THM, and other halogenated hydrocarbons which are considered carcinogenic for experimental animals.

A research into an alternative disinfection technology is required in order to replace the use of chlorine gas disinfection method and hence avoid the unnecessary risks and negative side effects listed above. A proven methodology introduced by LVPG International [9] is investigated in this research in order to check its applicability as a powerful disinfection technique that may be applied in Iraq to provide safe sterile potable drinking water. It is a base technology that has been available for several years. Such methodology existed in Germany since 1998 in Frankfurt airport. The process has always offered great potential but not been realized industrially.

The LVPG Generator makes use of an innovative "cell" device where the water/saline solution is brought into contact with the electrodes hence, producing two discrete solutions, one of which is acidic, the other is alkaline. AnoFluid is a highly effective oxidizing solution. The products have two different effects and may be added to water in a particular ratio, either manually or automatically using dosing pumps or water depending on the water attributes and the desired results [9].

The goals of this study are as follows:

- To use the LVPG Generator that uses only water, salt, and electricity to generate two solutions called AnoFluid and CathoFluid required in the sterilization and disinfecting process of drinking water.
- To collect several samples of water from different locations in Baghdad in order to test the level of contamination and hence test their suitability as drinking water.
- To develop a simulation model of the LVPG Generation process to be employed in this study. The simulation study is to be conducted using simulation software called ChemCad.
- Define the ratio of AnoFluid to water that performs the required percentage of contaminant reduction.

Experimental Set-Up and Procedure

The Experimental Set-Up

A pilot plant is set up to study the performance of the LVPG generator in producing a disinfectant solution (AnoFluid and CathoFluid). The purpose of the LVPG pilot plant study is to test a new technology for disinfector generation and hence observe performance of an alternative treatment process. The pilot study approach allowed detailed testing to proceed concurrently with update of critical drinking water regulations in Iraq and produce a set up for future requirements. The Pilot plant set up consists of the LVPG Generator, Softener system, Circulating pump, and AnoFluid, CathoFluid, water supply Tanks. These are connected with piping system as shown in Figure E1. The Generator is shown figures E2 – E4.

Figure E1: Pilot Plant front view

Figure E2: LVPG Generator

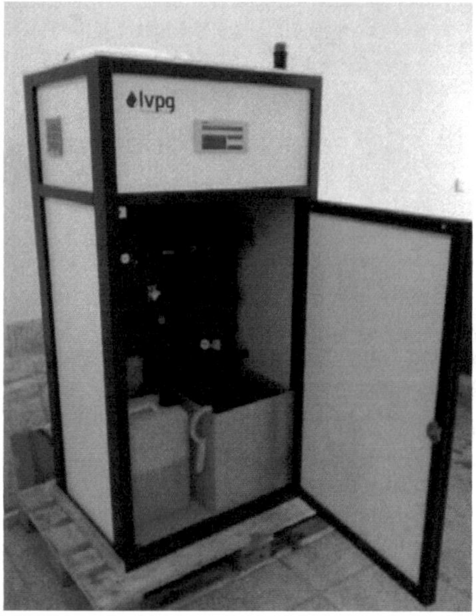

Figure E3: Inside the LVPG Generator Cabinet

Figure E4: Inside LVPG cabinet showing the AnoFluid / CathoFluid Generator cell, Automated "Cleaning" system, Redox monitoring loop, Input (water) pressure regulator, Saline solution mixing tank, Saline dosing system

Process description

Water characterization was conducted before passing water through filter and softener system indicated in the set-up diagram Figure E1. Water exiting the softener was checked to have 0dH (hardness=0), [9] i.e. must contain no salts otherwise cell will not function properly (a condition set by the manufacturing company). It is very important to constantly check hardness of water all throughout the test. 12 Kg of NaCl salt (DIN EN 973 of purity 99.9%) was mixed with about 44 L of softened water which produced the saturated saline solution. It is allowed to reach the recommended level in the saline solution tank. 2Kg of citric acid powder (CAS No 5949-29-1) is dissolved in 10 L of distilled water in the citric acid cleaning tank. It is then connected properly. Internal water pressure regulator (feeding pre-softened water) to the LVPG system is set at 3 bars. Water flow into system is adjusted to 100 l/h. The red Master Switch "On" is applied. There will be a system "Wait period" of approximately 120 min from "power on" to allow the salt water mixture to produce optimum saline solution mix. After the wait period F1 on the control panel screen is pressed to start the generation operation to produce AnoFluid / CathoFluid.

Samples Collection

Samples were collected from 7 locations in Baghdad. We have also managed to obtain a sample from Basra city as it is well known for its high salinity. We thought it would be interesting to make comparisons with it in this research. Sampling commenced two weeks before an initial "ramp up" period. AnoFluid samples were also collected and used in our tests. AnoFluids samples were used within 3 days of its production as it is known to dissociate into salt and water i.e. its original constituents when left un-used for more than 3 days.

Parameter	Al-Karkh project drinking water mg/l	Al-Rasheed project drinking water mg/l	Dijla project drinking water mg/l	Al-Karama project drinking water mg/l	Al-Kadisia project drinking water mg/l	Dijla project drinking water mg/l	Al-Basra project Drinking water mg/l
Major Ions							
Alkalinity	136	142	154	139	142	154	NP
Calcium	60	80	75	82	78	75	600
Chloride	48	89	84	86	93	84	800
Free residual Chlorine							
Conductivity	466	976	867	950	1039	867	
Flouride	0.14	0.13	0.11	0.19	0.13	0.11	0.77
Hardness	265	338	290	339	331	290	1000
Magnesium	27	34	35	32	33	35	300
Nitrate as N_2	10	10	20	10	20	25	16
Nitrites as N_2	M	M	M	M	M	M	<0.003
Nitrites as NO_2	0.001	0.001	0.001	0.002	0.001	0.001	M
pH	7.4	7.4	7.4	7.9	7.5	7.4	6.1
Sodium	250	300	400	350	400	360	400
Sulphate	135	222	220	258	223	220	500
TDS	307	560	505	474	539	505	2000
TSS	202	540	392	203	360	392	150
Appearance							
Color	<5	<5	<5	<5	<5	<5	10
Taste and odor	Acceptable	Acceptable	Acceptable	Acceptable	Acceptable	Acceptable	Acceptable
Turbidity	1.19	24	2.7	4.7	3.55	2.7	6 units
Total UV Transmittance	NP	NP	NP	NP	NP	NP	NP
Metals							
Aluminum	0.11	0.1	0.06	0.08	0.13	0.06	
Antimony	M		M	M	M	M	<0.005

Arsenic	0.02	0.05	0.03	U	0.005	0.04	<0.005
Barium	0.9	1	1	1	1	1.2	<0.01
Boron	M	M	M	M	M	M	
Cadmium	0.005	0.05	0.05	0.008	0.005	0.006	<0.003
Chromium	0.007	0.07	0.01	0.1	1	0.03	<0.02
Cyanide	0.01	0.05	0.1	0.4	0.08	0.01	
Copper	1.5	1	1.5	1.6	2	1.9	<0.01
Iron	0.05	0.45	0.08	0.45	0.09	0.08	
Lead	0.02	0.05	0.05	0.003	0.05	0.003	<0.01
Manganese	0.5	0.5	0.005	0.3	0.05	0.08	
Mercury	0.004	0.005	0.1	0.005	0.007	0.0009	<0.001
Nickel	0.05	0.07	0.5	0.09	0.7	0.008	
Selenium	0.04	0.02	0.3	0.1	0.08	0.04	<0.005
Thallium	U	U	0.008	U	U	M	
Uranium	U	U	U	U	U	U	
Zinc	M	4	2.5	M	U	U	2.5
Other Tests							
Dissolved Organic Carbon	0.03	0.06	0.05	0.009	0.5	NP	0.02
Total Organic Carbon	0.1	1	0.6	1	1	NP	0.6
Microbiological Parameters							
Total and Faecal Coliform	10	15	25	20	15	40	30
E.Coli	15	20	10	25	22	35	30

Experimental Procedure for AnoFluid demand

AnoFluid produced is tested to comply with the DIN standards: DN EN SO 7393 (G4-2) for Drinking Water. Once AnoFluid has been produced and stored in the appropriate volumes it is necessary to provide a mechanism to "add" the fluid in the correct ratio into the water to be treated. In order to determine the amount of AnoFluid to be added to drinking water for disinfection purposes, 0.5-1 mg/l residual chlorine was taken as the basis [10, 11]. It also depends upon the water quality specification and the measurable components as set out in the Iraqi Water specification and also based on experience from other users and on the literature [9]. The procedure carried out in order to determine the amount of AnoFluid actually added in order to satisfy the limit of residual chlorine is as follows:

- 1% AnoFluid solution is made up;
- 6 drops of solution is added to 1 liter of clean water;

- Free residual chlorine should be 5-1 mg/l, if not we either add more drops or dilute with clean water until this range is reached (this is original chlorine);
- 500 ml is measured into second container;
- 500 ml of water is added to be tested, mixed well and left for 30 minutes,
- water is tested for free residual chlorine by the usual testing procedures foe chemical constituents of water;
- Chlorine demand = original chlorine/2 – residual chlorine.

Results of the above experimental procedure indicates that AnoFluid should be added to water in the ratio of 1:100 (1 liter AnoFluid to 100 liters of water), making residual chlorine in treated water in the range of 0.5-1 mg/l. AnoFluid is added to all samples of water collected from different parts of the country in the same ratio.

Bacteriology Test Method

Relatively simple and inexpensive indicator methods are available for routine monitoring of the microbiological safety of water and the efficiency of treatment processes [11]. Most reliable results are obtained by high frequency testing for indicator organisms selected for particular purposes. Routine monitoring programs for drinking water may be based on tests for faecal streptococci, Thermotolerant Coliform organisms or Escherichia coli. Under certain circumstances, tests for the heterotrophic plate count and coliphages may be included. These tests are simple, inexpensive and yield results in a relatively short time.

Results and Discussion

Process Description

ChemCad 5.2 working under widows is a steady state chemical process simulator. A process simulator is a specialized calculator which enables us to sequence the equations that "model" unit operations in a useful way, specify enough variables to reduce the degree of freedom to zero, and then simultaneously solve a large number of interrelated equations for the unknown variables. What is special with this simulator is that it has a user friendly interface. It is easy to change equipment specifications and vary the composition of the influent water. The simulator runs much faster than the actual process. It is used to gain better understanding of complex process, and models obtained may be used to predict future plant performance. The results from the simulation model may give important guidelines to full scale plant design and operation [12].

In the simulation program figure (3), the generation proceeds in two stages. First stage, involves mixing water and NaCl in the saline water tank in which solution must be allowed to reach saturation limit before pumping saline solution into the second stage, which simulates the actual generation process taking place at the AnoFluid / CathoFluid generation cell.

ChemCad Simulation Software

Figure(3) ChemCad Simulation Flow Sheet

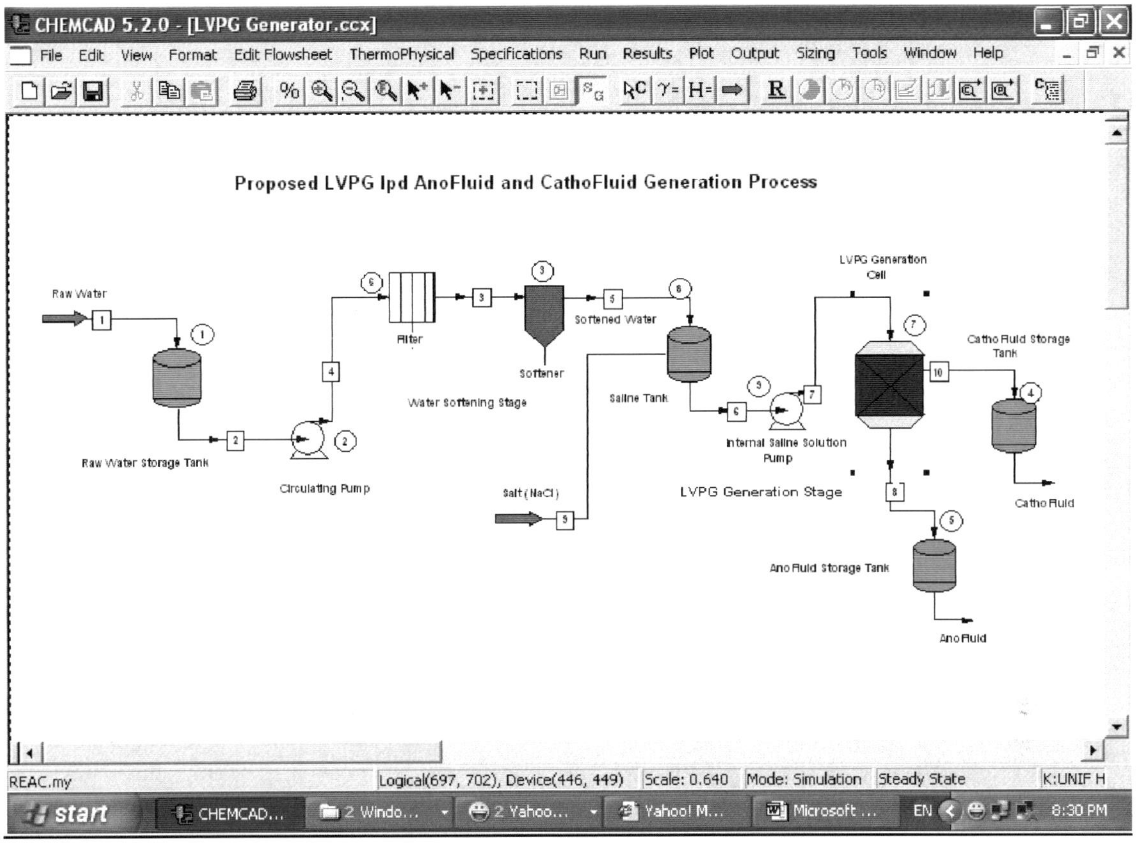

Table (1): AnoFluid and CathoFluid analysis results before addition to water

Characterization	**AnoFluid**	**CathoFluid**
pH value range	2.4 – 4.5	11 - 13
Redox value range (ORP)	1050 – 1150 mV	-800 to -900
Major constituents	Cl_2	NaOH
	HClO	H(X)
	OH^-	$H_2(X)$
	H_2O_2	OH^-
	HO_2	
	ClO_2	
	O_3H	

Table (2): AnoFluid CathoFluid Analysis results

AnoFluid Component	%	CathoFluid Components	%
Cl_2	51	NaOH	60
HOCl	8	OH	5
HCl	2	H_2O	35
OCl	6		
O_3	6		
O_3H	5		
H_2O_2	12		
ClO_2	7		
H_2O	3		

Table (3): Results of Disinfection with AnoFluid

Parameter	Results
Fecal coliform bacteria	No detectable fecal coliform bacteria per 100 ml
Escherichia coli	No detectable ***Escherichia coli*** per 100 ml
Total coliform bacteria (a) 1 sample in a 30 day period (b) more than 1 sample in a 30 day period	No detectable total coliform bacteria per 100 ml At least 90% of samples have no detectable total coliform bacteria per 100 ml and no sample has more than 10 total coliform bacteria per 100 ml

The possible application of mixed oxidants as a disinfectant was investigated thoroughly in this work. Although care was taken in the experiments to preserve the mixed oxidant (AnoFluid) solution to harvest the total potential oxidative power, and the possible synergistic effects of all the oxidants present, its effect significantly exceed that of chlorine. This was due to the fact that the AnoFluid and solution contained other strong oxidants such as ozone and hydrogen peroxide as shown in table (1).

Over the monitoring program, 7 samples were tested in order to determine the impact of disinfectant change on the observed bacteriological and chemical parameters. Results of this work provided valuable insight into the phenomena that are involved when introducing AnoFluid to drinking water samples.

(Acridine Orange Count) is a measurement based on a DNA stain of the number of organisms in water. This suggests that AnoFluid was more effective at killing microorganisms in drinking water than chlorine. In addition, no coliform bacteria were detected in drinking water during the duration of the trial as shown in Table (3). Results from batch experiments showed that the greatest reduction in the bacterial numbers took place within the first minute. In most cases both the AnoFluid and chlorine killed more than 99% of the bacteria present, however, with AnoFluid it happened during the first minute of exposure. Increased contact time did not significantly reduce the numbers of the bacteria any further. In all the batch tests the average percentage reduction in bacterial numbers were 99,55 and 97,84 for the AnoFluid and chlorine solution respectively after five minutes contact time. The overall percentage reduction in the continuous flow tests were 99,85 for the AnoFluid and 99,61 for the chlorine.

It was also observed that disinfection with chlorine is not instantaneous; time is required in order that any pathogens present in water are inactivated. With AnoFluid, however, it was observed that all kinds of bacteria were killed within the first 5 minuts at most.

In this study we have considered 'residual' chlorine when estimating amount of AnoFluid addition requirements to drinking water. It was observed that the level of 'residual' chlorine required varied slightly with type of water supply and local conditions. Chlorine residual was maintained at a level between 0.5-1 mg/l after adding 1 part of AnoFluid to 100 ml of water.

The suggested disinfection mechanism for AnoFluid is as follows: Passing of electrical current through solution creates an "Out Gassing", this results in the creation of millions (1-10m per cc) of charged oxidants. These are held in suspension in the fluid for extended periods (Meta Stable). Bubbles carry same charge and repel each other and hence get attracted to opposite polarity of harmful bacteria and viruses. They penetrate exterior membrane of bacteria/ virus by osmosis and release oxidants and destroy them. It actually acts in a similar way to the body's natural defense mechanism [9].

Conclusion

ChemCad 5.2 is considered as a useful tool to be used in this investigation in order to develop an ideal PC windows program for the design, development and optimization of a water treatment plant making use of the LVPG base technology. It will run on Windows platform (95, 98, 2000, NT and Windows XP) and will run quickly and with great stability.

Laboratory tests indicate that AnoFluid can be up to 300 times more effective in eradicating harmful contaminants, which are known to breed and flourish in the Bio-Film. AnoFluid's success is due to its high redox potential. When compared with chlorine, AnoFluid works more quickly and effectively.The LVPG technology is purported to present the following advantages over present chlorine disinfection method:

- The efficiency of the oxidants produced is, by mass, higher than that of chlorine;
- Oxidant mixtures can be produced on-site using only electricity, water and sodium chloride;
- Oxidant concentration and composition may be adjusted according to specific needs; and
- A residual may be maintained in product water.

Results obtained however indicate that it is possible, based on the Redox potential of the solution, to produce an AnoFluid of consistent quality. Reproducibility of the results was good and it can be expected that this aspect of the performance of the mixed oxidant generator (LVPG) could be applied on large scale.

References

1. World Health Organization (2002a). Water and Sanitation: Facts and Figures. [On-Line]. Available: http://www.who.int/water_sanitation_health/General/factsandfigures.htm (accessed 12-17-02).

2. Cameron, G.N., J.M. Symons, S.R. Spencer, and J.Y. Ma. 1989a. "Minimizing THM Formation during Control of the Asiatic Clam: A Comparison of Biocides." *J. AWWA*. 81(10):53-62.

3. Chick, H. 1908. "Investigation of the Laws of Disinfection." *J. Hygiene*. 8:92.

4. Hoigné J. and H. Bader. 1988. "The Formation of Trichloronitromethane (chloropicrin) and Chloroform in a Combined Ozonation/Chlorination Treatment of Drinking Water." *Water Resources*. 22 (3):313.

5. Krasner, S.W., M.J. McGuire, J.G. Jacangelo. 1989. "The Occurrence of Disinfection Byproducts in US Drinking Water." *J. AWWA*.81 (8):41-53.

6. Singer P.C. 1992. "Formation and Characterization of Disinfection Byproducts." Presented at the First International Conference on the Safety of Water Disinfection: Balancing Chemical and Microbial Risks.

7. Singer P.C. 1993. "Trihalomethanes and Other Byproducts Formed From the Chlorination of Drinking Water." National Academy of Engineering Symposium on Environmental Regulation: Accommodating Changes in Scientific, Technical, or Economic Information. Washington, D.C.

8. Singer P.C., and S.D. Chang. 1989. "Correlations Between Trihalomethanes and Total Organic Halides Formed During Water Treatment." *J. AWWA*. 81(8):61-65.

9. LVPG International [On-Line]. Available: www.lvpg-international.com. (Accessed 12-17-2010).

10. Snead, M.C., et al. 1980. *Benefits of Maintaining a Chlorine Residual in Water Supply Systems*. EPA 600/2-80-010.

11. World Health Organization (1997). "Guidelines for drinking water Quality" Volume 3[On-Line]. Available: http:// http://books.google.ca/books (accessed 1-17-2010).

12. Chemstations [On-Line]. Available: http://www.chemstations.net. (Accessed 1-1-2011).

Design of an Activated Sludge Simulation Plant for the Treatment of Wastewater Using ChemCAD Simulation Software

Dr. Hadeer K. Hamid, Water & Air Enterprises Ltd., Vancouver, BC. Canada

Summary

The role of ChemCAD simulation software in designing an activated sludge wastewater treatment plant is investigated in this paper. The simulation model represents a typical existing wastewater treatment plant in Iraq. The information required is obtained from directly measured data. Since the parameters are tracked on-line during varying operating conditions, the emphasis has been on the quality and quantity of the effluent. The designed simulation plant was found to actually reduce the effluent characteristics down to the limits set by the ministry of environment regulatory agencies and industry in Iraq. In this paper a basic model is used to validate the simulation results obtained from ChemCAD simulation software or the design of the activated sludge wastewater treatment plant, it is known as the Lawrence and McCarty model, which indicates the applicability of the ChemCAD simulation software in wastewater treatment design and application.

Introduction

Simulation has become an increasingly important tool in the study of wastewater treatment processes especially the activated sludge wastewater treatment plants due to its complexity and until now simulation technologies have not reached practical application [3]. Wastewater treatment plants generally perform biological treatment using microorganisms called activated sludge. They are used intensively in the treatment of municipal and industrial wastewater.

In Iraq, with the tendency towards using more advanced wastewater treatment technologies, it has become necessary to study alternative simulation techniques capable of handling the complexity and major control requirements in the operation and validation of wastewater treatment plants that currently exist and the new ones to be designed. Large expectations are placed on the ChemCAD simulation model using the activated sludge technology which may respond to these needs. It was not until the last few years before legislation on municipal wastewater treatment became effective in Iraq. The focus is on the reduction of the oxygen depletion problems through removal of the biodegradable organic compounds [5].

An activated - sludge reactor is a system in which pre-treatment sewage is aerated to promote the growth of bacteria (cells) that gradually consumes the organics in the sewage. The result is the development of cells acclimated to the particular mix of substances present in the sewage and a significant consumption of the organic material; the effluent is a mixture of water with drastically reduced BOD content and suspended cells. This mixture is then passed through a clarifier (setting Tank) where the solids (mostly cells, then called sludge) are separated from the water [1, 6].

The process considered is a real wastewater treatment plant that exists in Iraq designed to handle 10 MGD. The influent average flow Q is 37854 m^3/day and average organic BOD5 loads of 374 mg/L the daily variation of dry weather conditions are based on measured data from the plant. The average wastewater composition used in this study is shown in table (1).

Table (1) Wastewater influent composition in the treatment plant under study

Contaminants	Amounts (mg/L)	Explanation
BOD5	150	Substrate, organic, soluble, biodegradable material
TDS	600	Total Dissolved Solids (non-biodegradable)
FSS	80	Fixed Suspended Solids (non-biodegradable)
i-VSS	40	Inert Volatile Suspended Solids
a-VSS	200	Active Volatile Suspended Solids (heterotrophic biomass)
n-VSS	10	Volatile Suspended Solids for Nitrifies (autotrophic biomass)
Ammonia	40	
CO2	60	

The wastewater treatment plant employed is a contact stabilization process [2, 9] similar to that shown in Figure (1). It is characterized by a two – step aeration system. Aeration of short duration (1/2 to 2 hours) is provided in the contact tank (aeration) basin where raw or primary – settled wastewater is mixed with the activated sludge. The effluent from the aeration basin is then settled in a final settling tank. The settled activated sludge to be recycled from the final clarifier is drawn to a separate re-aeration in a stabilization basin (aerobic digester) for 3 to 8 hours of aeration time.

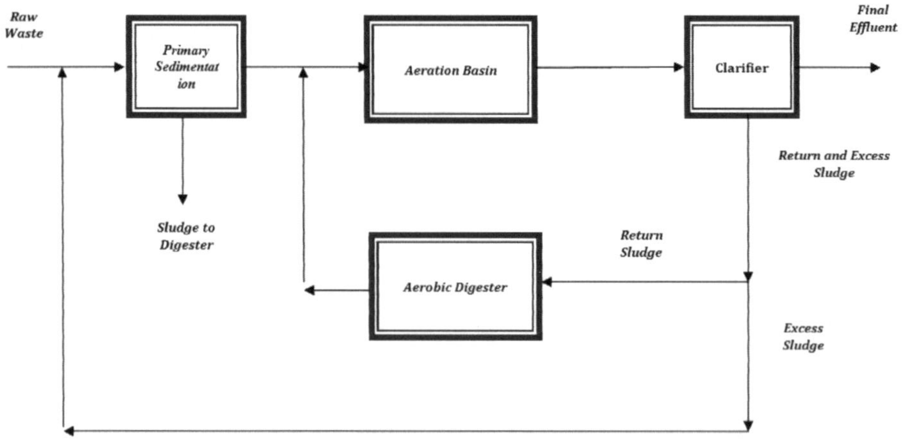

Figure (1) Contact Stabilization Activated Sludge Wastewater Treatment Block Diagram

It is then returned to the contact aeration basin for mixing with the incoming raw wastewater or primary settled effluent in addition to a short wastewater aeration time, the contact stabilization process has the advantage of being able to handle greater shock and toxic loadings than the conventional system [4].

Selecting the best mathematical model is the largest problem when designing any process. This paper focuses on the simulation of a real plant that exists in Iraq using ChemCAD simulation software. The validity of this technique is checked against a simple well-known model used extensively in the past for the design of activated sludge plants know as Lawrence and McCarty model [8].

Results and Discussion

ChemCAD [7] is a steady state chemical process simulator. A process simulator is a specialized calculator which enables you to sequence the equations that "model" unit operations in a useful way, specify enough variables to reduce the degree of freedom to zero, and then simultaneously solve a large number of interrelated equations for the unknown variables.

Steps to a simulation

The main steps used in the simulation are listed as follows:

- Start a new job (File/New)
- Select engineering units (Format/Engineering Units)
- Create flow sheet
- Select components (Thermo physical/Component List)
- Select thermodynamic options
- Define feed systems
- Input Equipment Parameters
- Run the simulation
- Review and output the results
- Validate the results

The process in figure (2) shows a simulation to the real municipal wastewater treatment plant. It consists of the following equipments divided between two main stages namely; the Biochemical Oxidation and the Sludge Treatment Process:

- ***Screen***: For the removal of coarse solids such as sticks, rags, cans, wooden materials, paper, textile and even floatable materials such as algae.
- ***Primary sedimentation tank***: The simulated design consists of a combination of a tank and component separator unit operations, the tank acts as a settling tank by offering a predesigned retention time hence allowing a removal of 40-60% of the suspended solids (FSS) and 30-50% of the BOD5.
- ***Aeration Basin***: The simulated design uses the Kinetic Reactor (CSTR) unit operation since it allows the user to enter the stoichiometric coefficients of the reaction and the rate of reaction equations. Here the influent is mixed with the activated sludge together with the oxygen supply allowing the aerobic process to commence and hence bacteria

start to grow faster performing the job of substrate utilization. The actual design retention time allowed is 0.5-2h.

- ***Clarifier***: The simulated design consists of a combination of a tank and component separator unit operations. Because the bacteria deplete the substrate, flocculation takes place; flocks of biomass will sediment gradually.
- ***Aerobic Digester***: The simulated design uses the Kinetic Reactor (CSTR) unit operation since it will perform a similar action to the aerobic digester, however, the retention time is 3-8h
- ***Thickener***: The simulated design consists of a combination of a tank and component separator unit operations.
- ***Primary and secondary vacuum filter***: Here Activated Sludge dewatering will take place.

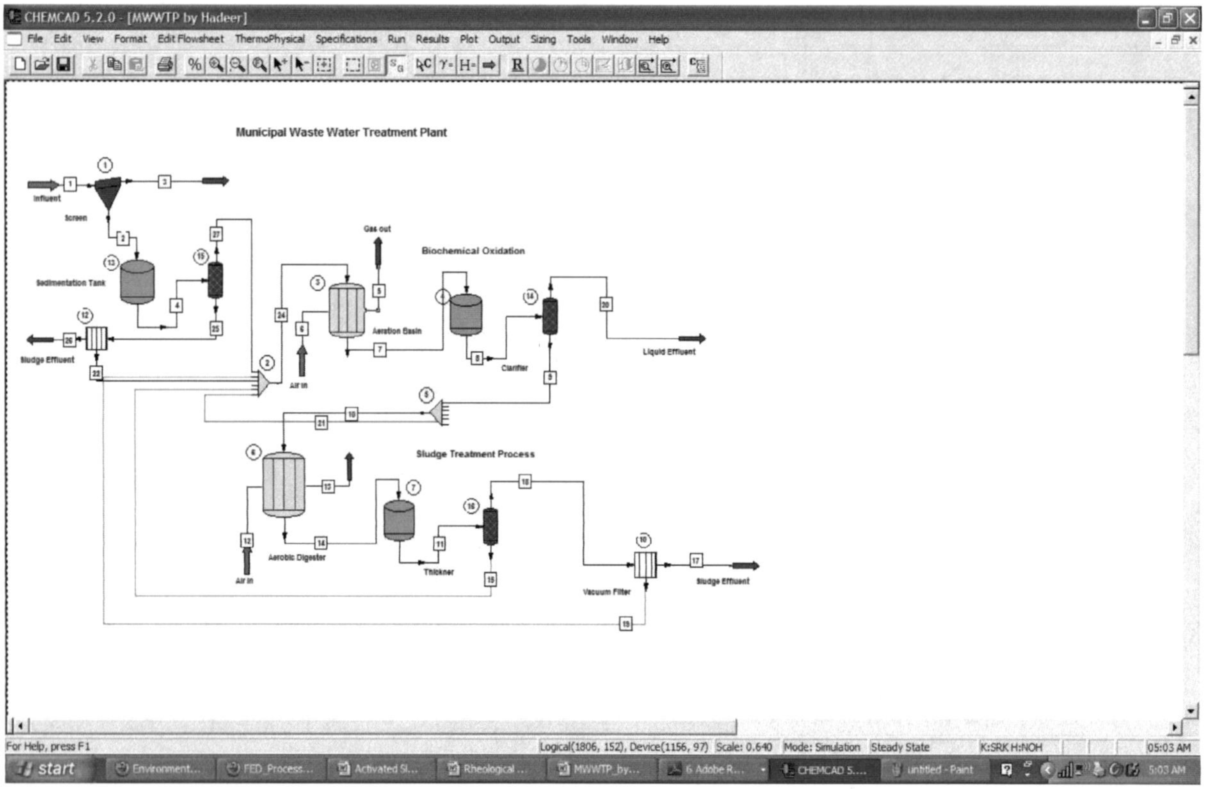

Figure (2) Simulation Flow sheet Using ChemCAD Software

The following table (2) gives the Activated Sludge Design Parameters adapted in this study based on data obtained from the real process.

Table (2) Design Parameters

Process Parameter	**Values**
BOD percent Removal	85-95
Q_d , hr	4-8
Organic loading, day^{-1}	0.25-0.50

Q_x , days	3-5
Sludge Recirculation Rate	0.25-0.5
Excess sludge production (mass of sludge per mass BOD destroyed)	0.5
Oxygen Requirement Minimum allowable dissolved oxygen mg/L Air flow rate, m^3/kg BOD	 0.5-1.0 90-100
Primary Sedimentation Tank 2 units Suspended solids removal Biological Oxygen Demand Removal Settled Solids Concentration Detention Time Overflow rate ***Dimensions*** Length, width, Depth	 40% 25% 4% 4 h 20 m^3/d-m^2 100m, 30m, 4m
Primary Vacuum filter 1 unit Cake Solid Concentration Filter Rate **Dimensions** Diameter, Length	 25% 20 kg/ m^2 –h 5m,6m
The Clarifier 4 units Suspended Solid Removal BOD Removal Settled Solids Concentration Detention Time Overflow Rate ***Dimensions*** Length, width, Depth	 40% 25% 4% 4 h 20 m^3/d-m^2 100m, 30m, 4m
The Thickener 2 units Detention Time Removes no additional BOD_5 Overflow Rate Dimensions	 4 h 20 m^3/d-m^2 100m, 30m, 5m
Secondary Vacuum Filter 2 units Cake Solids Concentration Filter Rate ***Dimensions*** Diameter, Length	 15% 20 kg/ m^2 –h 5m,6m
Activated Sludge Reactor 4units Detention Time Sludge Life Reactor Volume Recycle Ratio	 7 hr 120 hr 12500 m^3 0.3
Aeration Basin 4 units	

Detention Time	2 hr
Tank Volume	12500 m^3
Recycle Ratio	0.2

The main aim of the wastewater treatment plant is to maintain satisfactory concentration of organic and other pollutants. According to the new Iraqi regulations on the effluent of waste water treatment plant, the maximum concentrations in terms of biological oxygen demand (BOD), Suspended solids (SS), and total nitrogen (TN) are given by:

$$\mathbf{BOD^{max} = 25\ mg/l}$$

$$\mathbf{SS^{max} = 35\ mg/l}$$

$$\mathbf{TN^{max} = 10\ mg/l}$$

These constraints are to be satisfied during operation of the actual treatment plant.

Modeling Activated Sludge Processes

Activated sludge is the most commonly used biological wastewater treatment process. It is considered to be the most cost-effective way to remove the organic material from wastewater. Besides that, it is very flexible, reliable and has the capability of producing high quality effluent. Although the process was developed in the early 1900s, there has been a long transition between its development and the establishment of a theoretical framework that describes the process and provides a basis for its design and control. Only after 1960s, the process models started to shift from hypothesis to practically applicable models [3].

A model of the activated sludge process consists of three main things; a description of its biological and chemical "structure"; a description of kinetic relations; and a description of stoichiometric relations. The structure description is the components into which the substrate and the biomass are to be divided. The rate equation describe how rapidly one component is converted into another by various physical and biochemical processes are called the kinetic relations. And the stoichiometry description shows how much of a component is consumed or generated by a process. The starting point in modeling the process is to understand the biochemical operations during it. Regardless of the nature and complexity of the system involved, there are certain fundamental biochemical operations that occur universally in all systems. These are biological growth, decay, hydrolysis, ammonification, soluble microbial product formation, phosphorous uptake and release, and storage of organic compounds. The first three operations are the most common and considered in all activated sludge models [3, 8].

In this work a basic model is used to validate simulation results obtained from ChemCAD simulation software for the design of an activated sludge wastewater treatment plant. It is known as Lawrence and McCarty model [8]. There is no distinction between particulate and soluble COD in the influent. The nomenclature for modeling is defined in table (3), and a schematic idealization of the system is shown in Figure (3). In this system, the biomass was wasted daily in Q_w withdrawn directly from the reactor, and the volume wasted was replaced with filtered effluent

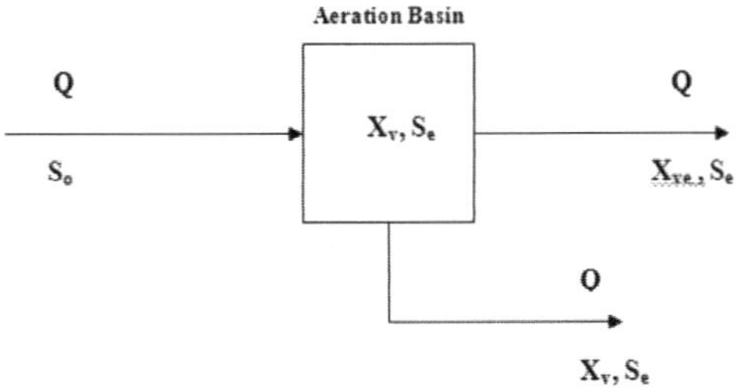

Figure (3): System schematic.

The steady-state mass balance for COD is

$$r_s = -\frac{S_0 - S_s}{\theta_d} \quad (1)$$

Biomass is taken as the total VSS in the reactor, and steady state mass balance for Xv is

$$r_{xv} = \frac{(Q - Q_w)X_{ve} + Q_w X_v}{V} \quad (2)$$

Earlier developments generally ignored a non biodegradable soluble organic matter component, but this modification is readily incorporated into simple models with significant improvements.

$$S_I = n_d S_o \quad (3)$$

And

$$S_e = S_s + S_I \quad (4)$$

Lawrence and McCarty used a Monod expression (Monod 1949) for substrate removal. The substrate removal model was first order with respect to biomass. At low substrate concentrations, hyperbolic expressions reduce to first-order expressions. The following substrate removal expressions were employed in the simple model.

$$r_s = -kS_s \quad (5)$$

$$r_s = -kX_vS_s \quad (6)$$

$$r_s = -\frac{kS_s}{K_s + S_s} \quad (7)$$

$$r_s = -\frac{kX_vS_s}{K_s + S_s} \quad (8)$$

The above expression and for any model, the unit of r_s are always expressed as milligrams of COD per liter per day (mg COD/L/d). The unit on k was adjusted to make each equation dimensionally consistent.

A net decay model is applied to biomass growth.

$$r_{xv} = -Y_H r_s - b_H X_v \qquad (9)$$

The units on r_{xv} for any model are milligrams of VSS COD per liter per day.

Equations (1) and (2) are used with equation (9) to determine the biomass concentration (X_v) as

$$X_v = \frac{\theta_x}{\theta_d} \frac{Y_H (S_0 - S_e)}{1 + b_H \theta_x} \qquad (10)$$

Lawrence and McCarty used the following coefficients in their model, these are: Y_H = 0.5 mg VSS COD/mg COD and b_H = 0.040 d^{-1}, n_d = 0.039, Ks = 98.3 mg/L, and k = 9.09 d^{-1}. An assumption is made here based on the effluent limit allowed in wastewater treatment literature [2].

Table (3) Definition of Symbols in the Model

b_H	= endogenous decay coefficient (d^{-1});
k	= substrate removal velocity coefficient (variable);
K_s	= substrate saturation coefficient (mg COD/L);
n_d	= inert soluble fraction of total influent (-);
Q	= influent flow rate (L/d);
Q_w	= waste flow rate (L/d);
r_s	= rate of substrate removal (mg COD/L/d);
r_{xv}	= rate of VSS production (mg VSS COD/L/d);
S_o	= influent total (degradable + inert) COD concentration (mg COD/L);
S_e	= effluent total (degradable +inert) soluble COD concentration (mg COD/L);
S_I	= inert soluble COD concentration (mg COD/L);
S_s	= effluent soluble degradable COD concentration (mg COD/L);
S_{so}	= influent soluble degradable COD concentration (mg COD/L);
X_{so}	= influent degradable particulate COD concentration (mg COD/L);
X_v	= VSS concentration in reactor (mg

	COD/L);
X_{ve}	= effluent VSS concentration (mg COD/L);
Y_H	= heterotrophic yield factor (-);
θ_d	= HRT (d);
θ_x	= SRT (d).
b_H	= endogenous decay coefficient (d^{-1});
k	= substrate removal velocity coefficient (variable);
K_s	= substrate saturation coefficient (mg COD/L);
n_d	= inert soluble fraction of total influent (-);
Q	= influent flow rate (L/d);
Q_w	= waste flow rate (L/d);
r_s	= rate of substrate removal (mg COD/L/d);

Stoichiometry in the aeration basin and aerobic digester used in this simulation

1. ***Substrate (BOD_5) Degradation***

BOD_5 + NH_3 + O_2 ===> a-vss + H_2O + CO_2

2. ***Nitrification***

NH_3 + O_2 + CO_2 ===> n-vss + H_2O + NO_3

3. **Biomass decay**

a-vss + O_2 ===> i-vss + FSS + H_2O + NH_3 + CO_2

n-vss + O_2 ===> i-vss + FSS + H_2O + NH_3 + CO_2

Model Validation

Table (4) presents data obtained at steady state in terms of COD in order to compare the applicability of ChemCad simulation software in wastewater treatment application when checked against the calculated results obtained by applying the Lawrence and McCarty model which has already been tested on a completely mixed activated sludge process and found to be successful in representing the process to an accepted level [9]. A factor of 1.42 was used for BOD conversion to COD.

Table (4): Results at steady state

COD Data	Lawrence and McCarty Model	ChemCAD Simulation Model
θ_x **(d)**	2.47	2.30
θ_d **(d)**	0.33	0.33
S_o **(mg/l)**	568	568
S_e **(mg/l)**	29.71	30.80
X_v **(mg/l)**	1833.37	1714.34

The designed simulation plant was found to actually reduce the effluent characteristics down to the limits set by the ministry of environment regulatory agencies and industry in Iraq Table (5).

Table (5) Effluent Components from the ChemCAD Simulation Software

COMPONENT	Effluent composition (mg/l)
Ammonia	4.3
CO_2	6.8
BOD_5	5.4
TDS	9.3
FSS	16.7
i-vss	1.6
a-vss	5.2
n-vss	0.5
NO_3	0.9
O_2	1.8
N_2	5

The results of the simulation are shown numerically in WinWord window as shown Table (6). The results show mass flow rates in each stream together with an overall input/ output balance.

Table (3) Simulation Results for an Activated Sludge Treatment Process.

Overall Mass Balance	kmol/h		kg/h	
	Input	Output	Input	Output
Ammonia	3.711	0.932	63.200	15.877
Nitrogen	0.000	0.400	0.000	11.210
Carbon Dioxide	2.154	12.154	94.800	534.900
Oxygen	62.502	61.961	2000.000	1982.680
Water	87601.224	87586.532	1578136.001	1577871.322
BOD5	237.000	64.790	237.000	64.790
TDS	948.000	1199.060	948.000	1199.060
FSS	126.000	123.030	126.000	123.030
i-vss	63.200	74.490	63.200	74.490
a-vss	316.000	11.640	316.000	11.640
n-vss	15.800	4.070	15.800	4.070
NO3	0.000	106.910	0.000	106.910
Total	89375.585	89246.000	1582000.006	1582000.00

CHEMCAD 5.2.0 Page 14

Job Name: MWWTP by Hadeer Date: 02/04/2011 Time: 15:28:26

FLOW SUMMARIES

Stream No.	1	2	3	4
Stream Name				
Temp C	20.0000	20.0000	20.0000	20.0000
Pres bar	1.0000	1.0000	1.0000	1.0000
Enth MJ/h	0.00000	0.00000	0.00000	0.00000
Vapor mole fraction	0.00000	0.00000	0.00000	0.00000
Total kmol/h	89313.0889	89313.0889	0.0000	89313.0889
Flowrates in kmol/h				
Ammonia	3.7109	3.7109	0.0000	3.7109

Nitrogen	0.0000	0.0000	0.0000	0.0000
Carbon Dioxide	2.1541	2.1541	0.0000	2.1541
Oxygen	0.0000	0.0000	0.0000	0.0000
Water	87601.2242	87601.2242	0.0000	87601.2242
BOD5	237.0000	237.0000	0.0000	237.0000
TDS	948.0000	948.0000	0.0000	948.0000
FSS	126.0000	126.0000	0.0000	126.0000
i-vss	63.2000	63.2000	0.0000	63.2000
a-vss	316.0000	316.0000	0.0000	316.0000
n-vss	15.8000	15.8000	0.0000	15.8000
NO3	0.0000	0.0000	0.0000	0.0000

Stream No.	5	6	7	8
Stream Name				
Temp C	20.0000	20.0000	20.0000	20.0000
Pres bar	1.0000	1.0000	1.0000	1.0000
Enth MJ/h	0.00000	0.00000	0.00000	0.00000
Vapor mole fraction	1.0000	1.0000	0.00000	0.00000
Total kmol/h	31.1809	31.2510	89035.5967	89115.5920
Flowrates in kmol/h				
Ammonia	0.0000	0.0000	2.9358	2.9358
Nitrogen	0.0000	0.0000	0.3070	0.3070
Carbon Dioxide	5.6551	0.0000	0.2343	0.2343
Oxygen	25.5258	31.2510	0.2156	0.2156
Water	0.0000	0.0000	87594.3424	87594.3424
BOD5	0.0000	0.0000	84.5000	84.5000
TDS	0.0000	0.0000	948.0000	948.0000
FSS	0.0000	0.0000	75.6000	75.6000
i-vss	0.0000	0.0000	2.1600	82.1600
a-vss	0.0000	0.0000	250.0000	250.0000
n-vss	0.0000	0.0000	13.5900	13.5900
NO3	0.0000	0.0000	63.7100	63.7100

Stream No.	9	10	11	12
Stream Name				
Temp C	20.0000	20.0000	20.0000	20.0000
Pres bar	1.0000	1.0000	1.0000	1.0000
Enth MJ/h	0.00000	0.00000	0.00000	0.00000
Vapor mole fraction	0.00000	0.00000	0.00000	1.0000
Total kmol/h	26065.2603	20851.6288	20943.0000	31.2510
Flowrates in kmol/h				
Ammonia	2.5771	2.0617	0.7815	0.0000
Nitrogen	0.0532	0.0425	0.0425	0.0000
Carbon Dioxide	0.0145	0.0116	0.2218	0.0000
Oxygen	0.1356	0.1084	0.1084	31.2510
Water	24601.2218	19680.9335	19674.8773	0.0000
BOD5	76.8200	61.4600	0.0000	0.0000
TDS	934.7800	747.8200	934.7800	0.0000
FSS	51.8600	41.4900	51.8600	0.0000
i-vss	79.8800	63.9000	102.8100	0.0000
a-vss	242.6100	194.0100	4.9200	0.0000
n-vss	12.8800	10.3040	4.6800	0.0000
NO3	62.4300	49.4900	167.9200	0.0000

Stream No.	13	14	15	17
Stream Name				
Temp C	30.0000	20.0000	20.0000	20.0000
Pres bar	1.0000	1.0000	1.0000	1.0000
Enth MJ/h	0.00000	0.00000	0.00000	0.00000
Vapor mole fraction	1.0000	0.00000	0.00000	0.00000
Total kmol/h	42.4799	20792.0000	6252.0749	1466.1211
Flowrates in kmol/h				
Ammonia	0.0000	0.7815	0.2343	0.5347
Nitrogen	0.0000	0.0425	0.0129	0.1189
Carbon Dioxide	6.1350	0.2218	0.0666	0.1204

Oxygen	36.3449	0.1084	0.0325	0.0013
Water	0.0000	19674.8773	5902.4646	50.4657
BOD5	0.0000	0.0000	0.0000	0.0000
TDS	0.0000	934.7800	249.6000	1184.4000
FSS	0.0000	51.8600	15.5600	46.3000
i-vss	0.0000	102.8100	30.8400	71.9700
a-vss	0.0000	4.9200	1.4800	3.4400
n-vss	0.0000	4.6800	1.4040	3.2800
NO3	0.0000	16.9200	50.3800	105.4900

Stream No.	18	19	20	21
Stream Name				
Temp C	20.0000	20.0000	20.0000	20.0000
Pres bar	1.0000	1.0000	1.0000	1.0000
Enth MJ/h	0.00000	0.00000	0.00000	0.00000
Vapor mole fraction	0.00000	0.00000	0.00000	0.00000
Total kmol/h	15190.1547	13725.1807	87570.3516	5204.0382
Flowrates in kmol/h				
Ammonia	0.5471	0.0123	0.3975	0.5138
Nitrogen	0.0296	0.0011	0.2813	0.0107
Carbon Dioxide	0.1552	0.0348	0.2436	0.0030
Oxygen	0.0759	0.0313	0.0888	0.0272
Water	13772.4172	13713.0506	87506.9124	4920.2334
BOD5	0.0000	0.0000	8.5100	15.3600
TDS	1184.4000	0.0000	14.6600	186.9500
FSS	36.3000	0.0000	26.3300	1.3700
i-vss	71.9700	0.0000	2.5200	15.9800
a-vss	3.4400	0.0000	8.2000	48.5200
n-vss	3.2800	0.0000	0.7900	2.5800
NO3	117.5400	12.0500	1.4200	12.4900

Stream No.	22	24	25	26
Stream Name				
Temp C	20.0000	20.0000	20.0000	20.0000
Pres bar	1.0000	1.0000	1.0000	1.0000
Enth MJ/h	0.00000	0.00000	0.00000	0.00000
Vapor mole fraction	0.00000	0.00000	0.00000	0.00000
Total kmol/h	54.8345	89135.0000	172.9500	135.8294
Flowrates in kmol/h				
Ammonia	0.0000	3.7109	0.0000	0.0000
Nitrogen	0.0000	0.0000	0.0000	0.0000
Carbon Dioxide	0.0000	2.1541	0.0000	0.0000
Oxygen	0.0000	0.0000	0.0000	0.0000
Water	51.8746	87589.7923	63.3000	29.1494
BOD5	2.9600	120.7100	59.2500	56.2800
TDS	0.0000	948.0000	0.0000	0.0000
FSS	0.0000	75.6000	50.4000	50.4000
i-vss	0.0000	63.2000	0.0000	0.0000
a-vss	0.0000	316.0000	0.0000	0.0000
n-vss	0.0000	15.8000	0.0000	0.0000
NO3	0.0000	0.0000	0.0000	0.0000

Stream No.	27
Stream Name	
Temp C	20.0000
Pres bar	1.0000
Enth MJ/h	0.00000
Vapor mole fraction	0.00000
Total kmol/h	89080.1551
Flowrates in kmol/h	
Ammonia	3.7109
Nitrogen	0.0000
Carbon Dioxide	2.1541
Oxygen	0.0000
Water	87538.0000

BOD5	117.7500
TDS	948.0000
FSS	75.6000
i-vss	63.2000
a-vss	316.0000
n-vss	15.8000
NO3	0.0000

Conclusion

ChemCAD simulation software is used in this study to simulate an existing Activated Sludge Wastewater treatment plant in Iraq. The simulated plant was capable of reducing the effluent streams down to the limit set by the ministry of environment regulatory agencies and industry in Iraq. It is concluded that ChemCAD may act as a tool that helps environmental engineers and engineering students in the design, operation and control of processes within environmental constraints.

In this work the Lawrence and McCarty model was chosen to test the validity of the activated sludge wastewater treatment simulation by ChemCAD, and therefore, check its applicability to different operating conditions. Results were found in excellent agreement indicating the applicability of the ChemCAD in the wastewater treatment application.

References

1. Nicholas P. Cheremisinoff, "Biotechnology for waste and wastewater treatment technology"; Noyes publications, New York, 1996.
2. Metcalf & Eddy, "Waste water engineering treatment, disposal and reuse"; McGrow-Hill, New york, 1991.
3. U. Jeppsson, "Modeling Aspects of Wastewater Treatment Processes", PhD thesis, Lund Institute of Technology, Lund, Sweden, 1996.
4. Z. Yuan et al., Wat. Sci. & Tech., 36, pp.81, 1997.
5. N.Padukone and G. Andrews, Wat. Res., 23, pp. 1535, 1989.
6. M. Henze et al., Wat. Res., 21, pp.505, 1987.
7. Chemstations [On-Line]. Available: http://www.chemstations.net. (Accessed 1-1-2011).
8. Lawrence, A. W., and McCarty, P. L. "unified basis for biological treatment design and operation." J. Saint. Eng. Div., Am. Soc. Civ. Eng., 96(3), 757-778.
9. Grady Jr., C. P. L., Daigger, G. T., and Lim, H. C. (1999). Biological wastewater treatment, 2nd Ed., Dekker, New York.

AUTHOR INDEX

AUTHOR INDEX

AIChE
120 Wall Street, FL 23
New York, NY 10005-4020

ISBN 978-1-61782-902-4

CURRAN ASSOCIATES INC.
proceedings.com